1290767l

Essays
in Paper Analysis

Essays
in Paper Analysis

Edited by Stephen Spector

Folger Books
Washington: The Folger Shakespeare Library
London and Toronto: Associated University Presses

Associated University Presses
440 Forsgate Drive
Cranbury, NJ 08512

Associated University Presses
25 Sicilian Avenue
London WC1A 2QH, England

Associated University Presses
2133 Royal Windsor Drive
Unit 1
Mississauga, Ontario
Canada L5J 1K5

The paper used in this publication meets the requirements
of the American National Standard for Permanence of Paper
for Printed Library Materials Z39.48-1984.

Library of Congress Cataloging-in-Publication Data
Main entry under title:

Essays in paper analysis.

"Folger books."
Bibliography: p.
Includes index.
1. Bibliography—Methodology—Addresses, essays,
lectures.   2. Paper—Analysis—Addresses, essays,
lectures.   3. Water-marks—Addresses, essays, lectures.
4. Manuscript dating—Addresses, essays, lectures.
I. Spector, Stephen, 1946–
Z1008.E86   1986   016.676'2   84-46121
ISBN 0-918016-87-8 (alk. paper)

Printed in the United States of America

Contents

Introduction

Stephen Spector

This book of essays, containing discussions of topics as diverse as medieval music manuscripts and bibliographic uses of nuclear physics, is intended to demonstrate some of the range and maturity of paper study. As the essays illustrate, this field has become increasingly sophisticated and complex. In this introduction, therefore, I attempt to contextualize the essays, and, in the process, to define some relevant terms, while noting some of the basic applications of paper study.

The Triumph of Paper

More than one medieval scribe doubtlessly preferred the use of a sturdy animal-skin writing surface to a comparatively flimsy sheet of paper. After all, parchment and vellum, chiefly sheepskin and calfskin respectively, provided, in the words of one modern scholar, "the most satisfactory material ever discovered for purposes of writing and illumination, tough enough for preservation to immemorial time, hard enough to bear thick strokes of pen or brush without the surface giving way, and yet fine enough for the most delicate ornamentation."[1] Paper, by contrast, seemed impermanent. Moreover, it bore the stigma of being perceived as an Islamic product, having been introduced into Europe by Arabs.[2] But the triumph of paper was inevitable, for paper is a *vegetable* substance. And as the demand for writing material increased in the late medieval period, and the papermaking industry grew, paper could be produced in greater quantities and at lower prices than animal skins. The quality of this paper has in many instances surpassed the bleak expectations of some skeptical medieval scribes. As Dard Hunter observes, the paper of the forty-two-line Gutenberg Bible, for example, has survived very satisfactorily this last half-millenium, and may well fare better in the next five hundred years than much of the handmade paper of the twentieth century.[3]

Papermaking is usually said to have been developed about A.D. 105 in China by T'sai Lun, who is thought to have substituted vegetable fibers for silk, the animal fiber that had earlier been used in making writing material. The Arabs evidently learned papermaking in 751 from Chinese prisoners in Samarkand, and paper is later noted in Baghdad and elsewhere in the Arab world. By the twelfth century, paper was manufactured in Xativa, Spain, and watermarked paper was being made by about 1282 in Frabriano, Italy.[4] But not until two centuries later did a paper mill operate in England, where John Tate of Hertford was providing paper for Wynkyn de Worde by around the year 1495.

Prior to this, and for nearly another two centuries afterward, most English white paper was imported, especially from France, and in particular Normandy.[5] The English tardiness in paper manufacture is understandable, since, as Philip Gaskell reminds us (p. 60), Englishmen wore wool, but white paper was made of linen rags. Indeed, the triumph of paper was made possible by the triumph of linen over wool in clothing, especially underwear, in Western Europe, since this made rags common enough to permit comparatively cheap papermaking.[6]

The hand fabrication of paper in the medieval period, or even today, has changed little since the invention of paper.[7] Essentially, the method involves placing a mould into a vat of macerated liquid paper pulp. The mould is then brought to the surface covered with the fiber, with the excess allowed to run off the far edge. The **deckle,** or frame around the mould, is removed, and the sheet, after drying briefly, is turned onto a felt of some kind. The mould is typically a wood frame without a top or bottom, with a surface in ancient times of woven cloth, then later of bamboo strips, and then, in medieval times, of metal wires. In each case, the surface held the pulp while letting the watery excess drain through. The metal wires of the mould consist of fine **laid wires** running parallel to the longer side of the mould, as well as **chains,** which are heavier, more widely spaced wires crossing them perpendicularly. Paper made on such a mould is called **laid paper,** which prevailed until about 1756, when **wove paper,** made on a mould with finely woven wire mesh, began to appear.

Watermarks

Relatively late in the history of papermaking, there was added a curious and somewhat mysterious aspect of the craft: the use of watermarks. Makers of ancient oriental paper did not use watermarks, perhaps in part because bamboo was unsuitable to the addition of designs.[8] Though metal wires were used in making paper in twelfth-century Spain, no watermarks seem to have been applied in Europe till more than a century later in Italy. Once they did appear, it is not always clear what they signified—or even why they came to be called "watermarks." Yet these marks became general, appearing in thousands of different designs, several dozen of which might appear in a single codex. For the scholar who is willing to shine lights on them or hold them up to the sun, measure and reproduce them, and, to use Allan Stevenson's phrase, stand on his head to make them out, these marks can be an extremely valuable source of information.

Watermarks are designs impressed on the paper by wires that were twisted into shapes and then sewed onto the mould. They do not appear to have originally indicated the size of paper.[9] Hunter says that they were perhaps nothing more than the mere fancy of the papermakers, and that "their ancient significance must remain more or less obscure."[10] Whatever their symbolic significance, Gaskell observes that the earliest marks were personal or trade marks of individual papermakers and mills.[11] It is unknown why these forms were called watermarks, and according to Hunter, this term does not appear until 1790, in an English patent. He considers the French *filigrane* and the Dutch *papiermerken* more suitable, and Stevenson says that *papermark* is a useful synonym in watermark studies, though it is ambiguous and master papermakers do not employ it.[12]

By the fifteenth century, most paper was watermarked, and a broad variety of designs had been employed.[13] Hunter (p. 300ff.) divides watermark images into four categories: those containing simple images, like crosses, ovals, circles, and hills; those presenting images of man and his works, ranging from hands, feet, and heads (including a fifteenth-century profile of a Negro slave) to keys, anchors, and horns; those with flora and nature images; and those with wild, domesticated and legendary animals, including bulls, fish, elephants, dogs, unicorns, and dragons.

In the sixteenth and seventeenth centuries, **countermarks** gained currency. These were often initials or other symbols that appeared opposite the principal watermark, on the other half of the mould. French countermarks indicated the papermaker's name until the end of the hand-press period, while English and Dutch countermarks lost their trade significance more quickly.[14]

Several collections of watermark tracings have been published, notably Charles Möise Briquet's *Les Filigranes* (1907). Briquet visited a hundred European manuscript archives, and traced and categorized according to subject, type, and date some sixty thousand watermarks from dated documents. Growing blind as he worked, he published 16,112 of them in the four-volume collection. This formidable work has been followed by several others, and, as G. Thomas Tanselle notes in his excellent article "The Bibliographical Description of Paper," students of English books should be especially aware of the collections of Edward Heawood, W. A. Churchill, and Alfred H. Shorter.[15] Among recent works, the *Findbücher* published by Gerhard Piccard from his watermark archive at the Hauptstaatsarchiv Stuttgart deserve special notice. For detailed reference to publications containing watermark reproductions, see Phillip Pulsiano's "Checklist of Books and Articles Containing Reproductions of Watermarks," in the present volume.

The Description of Watermarks

The description of a watermark should begin with the type of image, such as "Bull's Head Tau" or "Bunch of Grapes," preferably based on the categories in *Les Filigranes*, or on another collection of watermark reproductions if appropriate. If the mark closely resembles one or more of these tracings, this may be indicated by a reference such as "cf. Briquet 15161." If one should meet up with precisely the same mark in Briquet, this can be indicated by a notation such as "Briquet 15161." If no similar mark appears in the catalogues, one should note the closest category of image. A description of the mark is often useful as well.

After identifying the image, one should measure the height and breadth of the mark, including the position of the mark with respect to the chain lines. **Chain lines** are the ribs or grooves impressed on the paper by the chain wires of the paper mould. The side of the paper that faced down on these wires is called the **mould side,** and can usually be identified by shining light on the paper at various angles; the shadows that appear in the grooves reveal the chain lines, especially in the margins.[16] The **wire lines,** the indentations formed in the paper by the laid wires, though usually finer and much closer together than the chain lines, are also discernible in this way. A **center chain line** splits a watermark, while **attendant chain lines** (or **inner** and **outer chain lines**) surround the mark or split its edges. The chain and wire lines are in effect papermarks too, and even when no watermark image appears on a leaf, one may wish to measure the distances between the chains in order to distinguish paper types; this can be one effective way to identify cancels, for example. Paper without marks should be described as "unwatermarked" or "unmarked."

Allan Stevenson's system of measurement notes first the height and then the breadth of the watermark, with the latter showing the position of the chain lines relative to the mark. For example, a Bull's Head mark measuring 51 × 4[30] 7 is 51 mm high and 30 mm wide, with the left chain 4 mm from the mark and the right one 7 mm from it, on the other side of the image. The chains themselves are thus 41 mm apart. If the chains split the mark, this can be represented by colons or vertical lines within the square brackets. Thus 2 [15:7] 2 indicates an image split by a chain 15 mm from the left border of the image and 7 mm from the right border, and centered between two chains at distances of 2 mm on either side. A measurement of [5:15:2] indicates attendant chains at distances of 5 mm from the left and 2 mm from the right border of the image. Finally, measurements of 0 [10:5] 0 or [:10:5:] indicate that the edges of the mark sit exactly on chains; [10:5] will serve the same purpose if the omitted details are understood.[17]

Watermarks are normally measured with the mould side up and with the measurements recorded *away* from the **deckle edge** and *into* the center of the sheet.[18] The **sheet** refers to the paper as it is made on the mould. It is folded into **leaves,** and each side of a leaf is a **page.** One can help identify the direction toward the center of the sheet by considering the **format,** which indicates the number of leaves formed by the folding of the sheet. When the sheet was folded once, across the longer side, the result was the **folio** format, with two leaves and four pages. Watermarks were normally sewn in the center of one half of the oblong mould. As a result, when the sheet was folded in half, the mark would typically appear in roughly the center of one leaf, while the other leaf was unwatermarked.[19] The chain lines usually run vertically in this format, and the watermark on the marked leaf should be measured *toward* the folio fold, since that is the center of the sheet. When the sheet was folded a second time, across the first fold, the format was **quarto,** with four leaves and eight pages. The chain lines would now normally run horizontally, and the mark should be measured toward the top of the leaf.[20] Watermark measurement is easier in folios, since the image stands in the center of the leaf. In quartos, by contrast, the mark is typically divided roughly in half at the fold of two leaves, while the other two leaves are unmarked. Watermark measurement in this format must therefore be approximate, taking into account portions of the mark that may be lost in the fold. The designation mL, or mLF° (mould side left, or mould side left folio), is used to indicate that the mark sits on the left half of the sheet when seen from the mould side, while mR or mRF° places it on the right half.

Countermarks should be identified and measured similarly, and can be indicated by the term *countermark* or by a long equals sign preceding the measurement.

Twins, States, and Sewing Dots

Mills regularly used *pairs* of moulds in producing paper, and these companion moulds produced companion watermarks. The artisans who made the moulds, and sometimes the papermakers themselves, often created twin marks for a pair of moulds. These twins were "fraternal" rather than identical, but were sufficiently similar to have deceived many students of incunabula and Renaissance books. Very often, close examination shows that what appears to be a run of a single mark is in fact two, twins born together on companion moulds. "Watermarks like wrens," says Stevenson, "go in pairs," and he suspects that this fact may be helpful in unmasking frauds and forgers who complacently matched pot tops with pot bottoms without being aware of the differences between twins.[21] Descriptions of paper books should distinguish between watermark twins, noting differences, for example, in design, shape, or position. Stevenson offers ten points of possible difference:

1. Differences in *mould end:* one twin mark is centered in a different half of the mould than is the other, with one reading *in* and the other reading *out.*
2. Difference in *chain position:* the twins differ in their positions vis-à-vis the center chain or attendant chains.
3. Differences in *chain space* or *chain pattern:* differences of 2 mm. or more in the spaces between chains or the pattern of chains.

4. Difference in the *slant* of one of the marks with respect to the mould.
5. Difference in *reversed pattern:* one mark is the mirror-image of the other, except for names and initials.
6. Difference in *label:* one mark has, for example, a full name, while the other has only initials.
7. Difference in the *countermark.*
8. Difference in *distinctive detail:* the marks differ in some element of design, for example.
9. Difference in *sewing:* the wire used to sew marks to the mould often left **sewing dots,** or marks, on the paper, and these will differ in position in twin marks.[22]
10. Difference of *distortion:* parts of the design are bent out of position.[23]

Several of the fine differences between twin marks involve changes in a mark over time. Through use, a watermark gradually became distorted and worn, and wires could bend or break. Often the mark, or part of it, loosened and moved to its right on the mould (and therefore to its left on the sheet) and was resewn to the grid. With each such change in appearance or position, the watermark entered a different **state.** Stevenson declares that the realization that watermarks could go through a series of states was the foundation of modern paper analysis. The cornerstone of this approach is the certainty that one is following a single mark through its changing states, and to achieve this certainty, Stevenson depended in large part on the sewing dots. The number and position of dots would vary with each mark, providing, says Stevenson, a sure test of identity. He argues that the sewing dots constitute such powerful evidence that in certain instances he was able to recognize an identical mark in a very late state of distortion when only a few characteristic sewing dots remained.[24]

Watermark Reproduction

Detailed comparison between watermark twins, or similar marks, or among the states of a single mark, is made simpler and more reliable by comparing reproductions of the marks in question. Tracings like those in Briquet have limited value in this regard, since they often omit important information, such as sewing dots. More detailed and reliable alternatives now exist, however. One of these is beta-radiography, a method of obtaining a radiographic image of the papermark, while excluding most of the ink on the leaf. Other recent methods of watermark reproduction are the Ilkley technique, developed by Robin Alston, and the Dylux method, developed by Thomas L. Gravell. David Schoonover's essay, "Techniques of Reproducing Watermarks," in this book offers a detailed description of these techniques.

Many scholars are now employing these techniques, and libraries are beginning to form collections of watermark radiographs and photographs. Projects presently underway are also producing catalogues of watermark reproductions far more detailed and accurate than the tracings drawn with such dedication and labor by earlier generations of scholars.[25]

Analysis of Paper Evidence

Bibliography can be defined as the study of the material transmission of books. Since the primary material of most books is paper, the potential applications of paper evidence in book detective work are of course various, and will depend on the nature of the problem and the ingenuity of the investigator.

Paper has been an important source of information, for example, in cases involving forgery and other forms of misrepresentation. As early as 1771, James Whatman the younger displayed a sophisticated sense of bibliographical analysis when he said of documents in a forgery case, "they will differ in a wire or something."[26] In our own century, two celebrated bibliographical discoveries of misrepresentation depended, in very different ways, on paper evidence. The first was Sir Walter Greg's proof in 1908 that several Shakespearian and pseudo-Shakespearian quartos published by Thomas Pavier bore false publication dates. In what has been called his most spectacular discovery, Greg was able to show that these plays, which bore three different dates of publication, 1600, 1608, and 1619, were in fact all printed in the year 1619.[27] Greg's method was to combine typographical information with the paper evidence. Inspired by *Les Filigranes*, which had been published only the year before, Greg observed the appearance of the same watermarks in the paper of plays bearing all three dates. The quartos of the *Merchant of Venice, Lear,* and *Pericles*, for example, are dated 1600, 1608, and 1619 respectively, and yet two discrete kinds of paper are common to all three of these plays. Citing Briquet's statistics on the brief lifespans of most paper moulds, and the short-term use of most paper stocks, Greg argued that the appearance of a single make of paper in one play dated 1600, and in another dated 1619 would of itself suffice to call these dates in serious question. When we are faced not with one make, but with a number of distinct makes of paper, occurring in the plays of different dates, the difficulty in the way of accepting these dates as genuine is infinitely increased.[28]

Greg concluded that Pavier, for purposes of his own, falsified the earlier dates and released all of the quartos in 1619. His findings concerning the watermarks were strengthened by Stevenson, who found not only the same marks, but also the same twins, in quartos bearing all three dates. In addition, Stevenson found several marks in the Huntington Library set of the Pavier quartos to add to the twenty-seven Greg recorded, including what he took to be a 1608 watermark in *Sir John Oldcastle*, a play bearing the date 1600, and a 1617 (or 1619) watermark in a *Henry V* dated 1608.[29]

Another bibliographical discovery based largely on paper analysis was John Carter and Graham Pollard's demonstration in 1934 that a number of books by the Brownings, Ruskin, Dickens, and others were in fact forgeries. Of these, the star piece was a copy of Mrs. Browning's *Sonnets from the Portuguese* purportedly printed in Reading in 1847. For some time, this book had been suspected of being less than it pretended to be, and Carter and Pollard proved it to be a forgery by having the paper microscopically analyzed. The results showed that the paper contains a chemical wood not employed in English papermaking until 1874 or later. Others of the suspect pamphlets failed on these and similar grounds, allowing Carter and Pollard to conclude:

> The analysis of the paper has turned our vague suspicion into proved fact, and settled our doubts, not only of the Reading *Sonnets* and the three Ruskins, but of eighteen additional first editions hitherto unchallenged.[30]

Typographical as well as other considerations supported the conclusion of forgery, and the evidence ultimately pointed to the bibliographer Thomas J. Wise as the culprit.

The methods employed by Greg and Carter and Pollard were seminal. Greg's close comparison of the paper in many books inspired the close comparison of watermark states later employed by Stevenson. And Carter and Pollard's analysis of paper was the precurser of modern scientific analyses that are now being developed. Such techniques as particle-induced X-ray emission (PIXE) and energy-dispersive X-ray fluorescence (XRF) show promise for bibliographical analysis, as is described in David

Woodward's "Analysis of Paper and Ink in Early Maps: Opportunities and Realities" in this book.

Gatherings and Cancels

Another application of paper study lies in determining the original makeup of books and in identifying leaves that may have been added or deleted. Analysis of this kind involves establishing the constituents of the **gatherings,** which consist of sheets of paper folded and sewn together. Gatherings, also known as **quires,** could comprise one or more single sheets, or even half-sheets. In the folio format, as we have seen above, each sheet is folded in half; this results in a **bifolium** consisting of two **conjugate leaves,** one normally with the watermark, the other without a mark. In a folio book, therefore, a gathering consisting of three folded sheets would contain six leaves, with each watermarked leaf conjugate with an unmarked leaf. If we arbitrarily call the watermark "WM," and indicate unmarked leaves by dashes, such a gathering could be represented as follows:

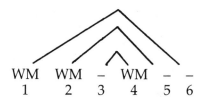

One can also look for patterns in the appearance of the mould side of each leaf. If, for example, the mould side appears on the recto of folio 1, it must also appear on the verso of the conjugate leaf, folio 6;[31] this must occur since these sides both faced down on the mould when the sheet was formed. Any irregularity in the watermark or mould-side distributions may indicate some disruption of the gathering.

In quartos, as noted above, each folded sheet produces four leaves, two of which divide the watermark, while two are unmarked. With our "WM" watermark, one leaf will carry the "W," the other the "M." With a watermark in its normal position centered in one half of the sheet, there were four possible patterns that could emerge from folding a whole sheet.

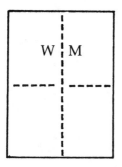

If, for example, the sheet were folded forward, then across the fold, the pattern would be – W M –. The other possible patterns were W – – M, M – – W, and – M W –.[32] In quarto books, gatherings often comprise single folded sheets falling in any of these patterns. An irregular scheme, such as W – – –, might indicate that the original fourth leaf, which bore the "M" portion of the mark, has been replaced by a **cancel** (also known as a **cancellans**). Other irregularities in the paper, such as differences in chain spaces, may also indicate cancellation.

Often, complex watermark distributions within gatherings can be shown to be mere elaborations of the four simple patterns noted above. If, for example, two sheets sitting one atop the other are folded together, a doubled version of W – – M may result, giving W W – – – – M M. Such an arrangement in fact appears in the second gathering of the *Castle of Perseverance*, a fifteenth-century morality play in Folger MS V a 354, where we find the following scheme (note that the leaves, misnumbered in the manuscript, have been shuffled here to reflect their original order):

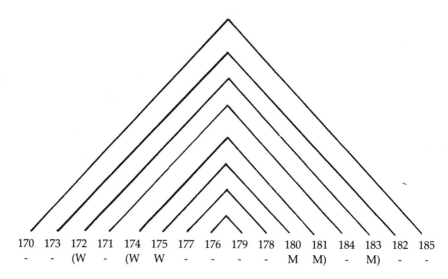

170 173 172 171 174 175 177 176 179 178 180 181 184 183 182 185
 -   - (W - (W W   -   -   -   -   M M) - M) -   -

Here folios 174 to 181 are two sheets illustrating the doubled W – – M pattern. If we look "beneath" these sheets by pretending to remove them from the gathering, we find the W – – M pattern in 172, 171, 184, and 183. But the first two and final two leaves of the gathering do not conform to any of the four watermark patterns. As we have seen, one watermarked leaf must appear for each unmarked one in these patterns, yet these four leaves are all unmarked. If whole sheets were used here, as seems to have been the case in the rest of this gathering and elsewhere in the codex, then four watermarked leaves, cognate with the unmarked folios 170, 173, 182, and 185, appear to be missing. And, in fact, their absence corresponds to breaks in the text. By comparing the banns preceding the play with the textual breaks, we can conjecture the contents of the lost leaves.[33]

Dating on the Basis of Watermarks

Greg's analysis of the Shakespeare quartos provided an early model for dating texts through comparing the watermarks, along with other evidence, in several books. Today, this comparative method of dating has become a highly sophisticated art, though its reliability has been the subject of some controversy.

The debate about dating on the basis of watermarks crystallized in the exchanges between two distinguished scholars, Allan Stevenson and Curt Bühler. Writing in 1957, Bühler observed that *Les Filigranes* is a work to which many students turn daily with gratitude, but he cautioned that "no one is quite certain for how long any particular mould could be used" and that "it is not clear how successful the methods for speedy distribution were."[34] He warned against depending too much on average life-spans, whether of moulds or paper stocks, and concluded that watermarks "furnish the student of fifteenth-century books with an additional (and important) tool for the dating of an incunabulum 'sine ulla nota,' possibly within a score or so of years as Briquet intimated."[35]

Stevenson, having earlier insisted on the importance of distinguishing fine points of watermark differentiation, wrote in "Paper as Bibliographical Evidence" (1962) that most scholars overestimate the value of simply comparing the watermarks in one's text with the tracings in Briquet. It would be far more instructive, said Stevenson, to follow a pair of marks through several books, watching the mould and mark deteriorate and grow old. In his Principle of Runs and Remnants, Stevenson addressed the question of the time lag between paper fabrication and use. He contended that paper purchased for use in manuscripts might last over a period of years, especially if the paper were of large size or high quality. With *printed* books, by contrast, paper was usually bought for printing a particular book, probably not long after manufacture, and was often used in *runs* over many gatherings. The *remnants* of a paper stock, however, could appear randomly in books years after the bulk of that paper had been used. One such remnant, said Stevenson, is the Pot watermark dated 1598 that turns up in the second quarto of *Hamlet*, dated 1604 or 1605.[36]

In the same essay, Stevenson offered an analysis of the *Missale speciale* to illustrate his thesis that by tracing the "life history" of the mark in the paper of several books, one can date with some precision (a fuller discussion, including typographical and other corroboratory evidence, appeared in his *Problem of the Missale speciale* [London, 1967]). Bühler had supported the idea that the *Missale* was printed around 1450. In 1960, however, both Gerhard Piccard and Theo Gerardy had used watermarks to reinforce earlier challenges to the mid-fifteenth-century date, noting that the paper of this missal appeared in books of the 1470s.[37] Stevenson extended this work by finding eleven books containing watermarks identical to ones in the *Missale* and then organizing these marks into their states. By locating the states of the marks in the *Missale* with reference to the other books, Stevenson aimed at a precise date for the *Missale*. He concluded that it was printed in 1473, possibly beginning in February and ending in September or October.[38]

In 1973, Bühler pressed further his case against relying uncritically on watermark evidence in establishing dates for books "sine nota." Acknowledging the great refinements in paper study in the preceding half-century, Bühler argued that while such methods can show the order of paper manufacture, they do not necessarily point to the order of paper use. He questioned whether in fact moulds always had a limited life, or were always used continuously; whether all the paper produced by a mould was necessarily used up in two or three years; whether it was used by the printer in the same order in which it came from the mould; and whether or not stocks of paper were ever accumulated.[39]

Arguments of this kind underscore the importance of employing discretion as well as precision in dating on the basis of watermarks. And they reinforce Stevenson's own assertion that watermarks provide but one form of evidence, and should be coordinated with the study of ink, bindings, and other relevant information. One must also keep in mind the distinction between runs and remnants, between large moulds and small, and between averages and particulars. Most important, consideration of the evidence within and among texts must be exact. Properly applied in this way, the comparative method can produce compelling results.

Indeed, analyses of the states of marks and moulds have been employed in interesting ways in recent years. One example of this is the debate concerning the date of the Mainz *Catholicon*, which bears a **colophon** stating that it was printed in 1460.[40] The book comes down to us in three impressions, worked on three presses, employing three distinct paper-stock groups. Studies by Gerardy, and by Eva Ziesche and Dierk Schnitger, have indicated that the three paper stocks of the *Catholicon* were made at three distinct times over a period of several years. They conclude, in fact, that most of the paper found in the *Catholicon* was not yet made in 1460; some seems to date from a decade or so later.[41] Paul Needham notes, however, that other evidence, including an inscription on one copy of the *Catholicon*, indicates that it must have existed by 1465, a date apparently too early for the paper evidence. He adds that the states of the type in the *Catholicon* and other books from the same press seem to predate indulgences of 1462 and 1464. Needham contradicts Ziesche and Schnitger's proposed solution, that there were three separate editions of varying dates, by pointing out that all three impressions represent the same typesetting throughout. His own solution is that the *Catholicon* printer used two-line slugs rather than movable types, printing the book at three separate times with three different paper supplies, but with the same slugs.[42] In the essay "Was the Mainz *Catholicon* Printed in 1460?" in the present volume, Curt Bühler addresses some of the anomalies concerning the *Catholicon*, and raises further commonsense questions.[43]

Dating through paper evidence is simpler when the watermark image itself contains a date, but here too cautions apply. Questions of time lag between production and use are of course relevant, and one must also be alert to the possibility that, as Dard Hunter says, a dated mould might have been used for many years with the same date (*Papermaking*, p. 294). Jan LaRue, in an article written with J. S. G. Simmons, observes that some French papermakers misinterpreted an edict requiring dates in watermarks from 1742 to mean that this same date should be used in later years as well.[44] Hilton Kelliher in his essay, "Early Dated Watermarks in English Papers: a Cautionary Note," in the present book records a similar phenomenon in England: an act encouraged English papermakers to include dates in watermarks as of 1794, and this date appeared in moulds until at least the turn of the century. And Hunter notes that the American paper manufacturer Joseph Willcox found it convenient to use a mould dated 1810 to produce paper nearly fifty years later (p. 294). In addition to the occasional appearance of dated marks on paper made after the year recorded, Kelliher warns in his article of an unexpected phenomenon: a date in a mark that is actually later than the text itself.

The Present Volume

This volume brings together essays relating both to paper study per se and to applications of paper evidence in various disciplines. In addition to the essays already mentioned, by Bühler, Kelliher, Pulsiano, Schoonover, and Woodward, it includes three essays on paper study and music scholarship: Frederick Hudson reviews this field in "Musicology and Paper Study—A Survey and Evaluation"; John Nádas provides a model description of a medieval paper music manuscript in "The Reina Codex Revisited"; and Alan Tyson examines the evidence in the paper of a Beethoven sketchbook in "Beethoven's Leonore Sketchbook (Mendelssohn 15): Problems of Reconstruction and of Chronology." In addition, the volume contains William Proctor Williams's "Paper as Evidence: The Utility of the Study of Paper for Seventeenth-Century English Literary Scholarship," which discusses the study of paper in seventeenth-century texts, with particular application to recently rediscovered manuscripts. The essays collected here evidence much of the vitality and growth of paper study in recent years, and confirm Allan Stevenson's assertion of over twenty years ago that

> Surely the time has finally come when paper may make a more serious and vital contribution to the unsnarling of bibliographical problems, to the art of the book detective.[45]

Notes

Among the many who generously offered advice and encouragement during the preparation of this volume of essays, John Bidwell, Warner Barnes, David Erdman, Leo Treitler, Richard Kramer, and, not least, the contributors themselves have my special thanks and appreciation.

1. Falconer Maden, *Books in Manuscript* (London, 1893), p. 9. G. S. Ivy, in "The Bibliography of the Manuscript Book," *The English Library before 1700*, eds. Francis Wormald and C. E. Wright (London, 1958), p. 35, cites recipes for making parchment and vellum. The word *parchment* derives from "Pergamum," the name of a city in Asia Minor. *Paper* comes from the Latin *papyrus*, though the two substances are very different.

2. Allan Stevenson, *The Problem of the Missale speciale* (London, 1967), p. 49; André Blum, *On the Origin of Paper*, trans. Harry Miller Lydenberg (New York, 1934), pp. 34–38. Blum (pp. 26 ff.) notes the role of Jews in the early paper industry, and includes its "Judeo-Arabic origin" (p. 34) as one of the bases for the medieval prejudice against paper.

3. Dard Hunter, *Papermaking through Eighteen Centuries* (New York, 1930), p. 297.

4. Cf. Blum, *Origin*, p. 22; Stevenson, *Problem*, p. 278.

5. Philip Gaskell, *A New Introduction to Bibliography* (Oxford, 1972, reprinted 1974), p. 60; Allan Stevenson, editor, *Supplementary Material Contributed by a Number of Scholars to C. M. Briquet, Les Filigranes* (Amsterdam: Paper Publications Society, 1976), p. 34.

6. Blum, *Origin*, p. 35; Stevenson, *Problem*, p. 49.

7. Hunter, *Papermaking*, p. 110.

8. Ibid., p. 289.

9. Ibid., p. 292.

10. Ibid., pp. 288, 289.

11. Gaskell, *A New Introduction*, p. 61.

12. Hunter, *Papermaking*, p. 293; Stevenson, "Paper as Bibliographical Evi-

dence," *Library,* 5th ser., 17 (1962): 197 n.4. The word *filigrane* comes from the Latin *filium* "thread, wire" and *granum* "grain."

13. Gaskell (*A New Introduction,* p. 61), notes, however, that poor-quality printing paper during the later hand-press period was often unwatermarked.

14. Gaskell, *A New Introduction,* p. 62.

15. G. Thomas Tanselle, "The Bibliographical Description of Paper," *Studies in Bibliography,* 24 (1971): 51–52. Tanselle notes that *Les Filigranes* does not extend beyond 1600, and does not cover Spain, Portugal, Scandinavia, or Britain. He therefore suggests that the student of English books should refer in particular to the following collections in addition to Briquet:

> Edward Heawood, "Sources of Early English Paper-Supply," *Library,* 4th ser., 10 (1929–30): 282–307, 427–54; "Papers Used in England after 1600," 11 (1930–31), 263–99, 466–98; "Further Notes on Paper Used in England after 1600," 5th ser., 2 (1947–48), 119–49; 3 (1948–49), 141–42; *Watermarks, Mainly of the 17th and 18th Centuries* (Hilversum: Paper Publications Society, 1950, reprinted 1957).
>
> W. A. Churchill, *Watermarks in Paper in Holland, England, France, etc., in the XVII and XVIII Centuries and Their Interconnection* (Amsterdam, 1935, reprinted 1965, 1967).
>
> Alfred H. Shorter, *Paper Mills and Paper Makers in England, 1495–1800* (Hilversum: Paper Publications Society, 1957).

Several other collections are noted by Tanselle (p. 52) and Stevenson, "Paper as Bibliographical Evidence," p. 198.

16. The other side of the mould side of the paper is the **felt side.** Tanselle ("Bibliographical Description," p. 34) notes that chain lines in *machine-made* paper are not a natural result of the manufacturing process. Rather, they are a design impressed upon the paper, and so are valueless for analysis.

17. Tanselle ("Bibliographical Description") says that a full bibliographical description of paper in books should estimate the size of the sheet and note whether the paper is laid (since all paper before the mid-eighteenth century was laid, this notation is actually unnecessary for books from before this time). If the paper is laid, the distance between the chains should be noted. Then the watermark and countermark (if one is present) should be identified and measured. Tanselle says that in addition to the dimensions and markings of the paper, one should test for thickness, color, and finish, recording only information that is of primary interest to the readers of the bibliography.

18. By metonymy, **deckle** means the rough edges of the sheet as well as the rim placed on the mould to prevent the macerated fiber from overflowing. If one chooses to measure *toward* the deckle edge, one should follow the measurement with a dash and the word *out.*

19. Gaskell (*A New Introduction,* p. 61) observes that "in the early days of paper-making the marks had been placed almost anywhere on the surface of the mould, but by the fifteenth century they were normally put in the centre of one half of the oblong." Cf. Tanselle, "Bibliographical Description," pp. 33–34.

20. Other formats include octavo, duodecimo, long twelves, and sixteenmo. **Turned chain lines,** which appear to run in the wrong direction for a particular format, occasionally show up, mostly in paper of the late seventeenth and early eighteenth centuries. See Gaskell, *A New Introduction,* pp. 80 ff.

21. Allan Stevenson, "Watermarks are Twins," *Studies in Bibliography* 4 (1951–52): 88, 90.

22. Stevenson notes that *sewing dots* were recorded in the first English books showing watermark tracings, Samuel Leigh Sotheby's *The Typography of the Fifteenth Century* (1845) and *Principia typographica* (1858); see Stevenson, "Paper as Bibliographical Evidence," p. 203.

23. List from Stevenson, "Watermarks are Twins," pp. 65–68.

24. Stevenson, "Paper as Bibliographical Evidence," p. 203.

25. See, for example, the catalogues of Thomas L. Gravell and George Miller listed in Phillip Pulsiano's "Checklist" in this anthology.

26. T. Balston, *James Whatman Father and Son* (London, 1957), p. 148, cited in Gaskell, *A New Introduction,* p. 62. In 1778, however, one John Mathieson forged banknotes so successfully that no difference could be detected between the actual watermark and his copy. He offered to reveal his secret in exchange for his life, but this offer met with an appalling shortage of bibliographical curiosity, and was rejected (see Hunter, *Papermaking,* pp. 263–64).

27. See F. P. Wilson, The Bibliographical Society, *Studies in Retrospect* (London, 1945), pp. 78–80.

28. W. W. Greg, "On Certain False Dates in Shakespearian Quartos," *Library,* 2d ser., 9 (1908): 122–23.

29. Allan Stevenson, "Shakespearian Dated Watermarks," *Studies in Bibliography* 4 (1951–52): 159–64.

30. John Carter and Graham Pollard, *An Enquiry into the Nature of Certain Nineteenth Century Pamphlets* (London, 1934; reprinted, New York, 1971), p. 55. For another fine detection of forgery, see Allen T. Hazen, *A Bibliography of the Strawberry Hill Press* (New Haven, 1942).

31. The **recto** is the front side of the leaf, which is on the right side of the opening of the book. The **verso,** the back of the leaf, is on the left side of the opening.

32. For a more detailed discussion of watermark distribution, see Stephen Spector, "Symmetry in Watermark Sequences," *Studies in Bibliography* 31 (1978): 162–78.

33. See Stephen Spector, "Paper Evidence and the Genesis of the Macro Plays," *Mediaevalia* 5 (1979): 224–27. The actual watermark in this gathering is a circle surrounding three mounds (cf. Briquet 11849, 11850).

34. Curt F. Bühler, "Watermarks and the Dates of Fifteenth-Century Books," *Studies in Bibliography* 9 (1957): 217, 219.

35. Ibid., p. 224.

36. Stevenson, "Paper as Bibliographical Evidence," p. 202.

37. Gerhard Piccard, "Die Datierung des Missale speciale (Constantiense) durch seine Papiermarken," *Borsenblatt* 16 (1960): 259–72, and *Archiv für Geschichte des Buchwesens* 2 (1960): 571–84; Theo Gerardy, "Die Wasserzeichen des mit Gutenbergs kleiner Psaltertype gedruckten Missale speciale," *Papiergeschichte* 10 (1960): 13–22, and "Zur Datierung des mit Gutenbergs kleiner Psaltertype gedrucktren Missale speciale," *Borsenblatt* 19 (1963): 884–92, also published in *Archiv für Geschichte des Buchwesens* 5 (1963), cols. 399–415.

38. Stevenson cited scholarship dismissing Eugène Misset's claim that the *Missale* "was printed before September 1468, because it did not include the Office for the Presentation of the Virgin, which was required at Constance from that time" (*Problem,* p. 154). In 1972, Bühler defended Misset's thesis in his "The *Missale speciale* and the Feast of the Presentation of the Blessed Virgin Mary," *Papers of the Bibliographical Society of America* 66 (1972): 1–11.

39. Curt F. Bühler, "Last Words on Watermarks," *Publications of the Bibliographical Society of America* 67 (1973): 1–16.

40. A **colophon** is an inscription or device normally placed at the end of a book or manuscript and containing such information as the title, the printer's or scribe's name, and the place or date of printing or transcription. For a reproduction of the Mainz *Catholicon* colphon, see Curt F. Bühler's "Was the Mainz *Catholicon* Printed in 1460?" in the present volume.

41. Theo Gerardy, "Wann wurde das Catholicon mit der Schluss-Schrift von 1460 (GW 3182) wirklich gedruckt?" *Gutenberg Jahrbuch,* 1973, pp. 105–25; and "Die Datierung zweier Drucke in der Catholicontype (H 1425 und H 5803)," *Gutenberg Jahrbuch,* 1980, pp. 30–37; Eva Ziesche and Dierk Schnitger, "Elektronen-radiographische Untersuchungen der Wasserzeichen des Mainzer Catholicon von 1460," *Archiv für Geschichte des Buchwesens,* 21 (1980), cols. 1303–50.

42. Paul Needham, "Johann Gutenberg and the Catholicon Press," *Papers of the Bibliographical Society of America* 76 (1982): 395–456. Needham (417ff.) rejects Piccard's argument that the *Catholicon* was indeed printed in 1460, as well as Gerardy's dating in the latter half of that decade.

43. Dr. Bühler's essay was written prior to the appearance of Dr. Needham's, and was revised to take into account Needham's conclusions.

44. Jan LaRue, with J. S. G. Simmons, "Watermarks," in the *New Grove Dictionary of Music and Musicians,* vol. 20 (London, Washington, D.C., Hong Kong, 1980); cf. Curt F. Bühler, "Watermarks and the Dates of Fifteenth-Century Books," p. 224 n.34; Stevenson, *Problem,* p. 98.

45. "Paper as Bibliographical Evidence," p. 212.

Essays
in Paper Analysis

1

Was the Mainz *Catholicon* Printed in 1460?

Curt F. Bühler

The earliest printings in Mainz, as is common knowledge, offer the student of early printing a variety of difficult problems.[1] An exception to this rule was apparently the first edition of the *Catholicon*.[2] The author of this work, a thirteenth-century Joannes Balbus of Genoa, enjoyed a considerable reputation for his scholarship. The colophon proclaims its place of printing (Mainz) and sets forth the year of production (1460), as the reproduction of this colophon makes evident.[3] There is every reason to believe that the printer of this *Catholicon* was the great inventor of the art, Johann Gutenberg, and this is now generally admitted to be a fact.

The many bibliographical references may be found through the descriptions provided by Margaret Stillwell and Hans Widmann.[4] On the strength of the findings of many generations of bibliographers, the imprint was generally assumed to be: Mainz [Johann Gutenberg?] 1460. There the matter rested till 1971.

In that year, Theo Gerardy set forth his belief that the evidence of the watermarks found in the book showed that the *Catholicon* had not been printed in the year asserted by the colophon.[5] He maintained (p. 23) that the year of printing must have been 1469.[6] This was a quite impossible date, since Gerardy judged that Gutenberg had been the printer—but the inventor had died before this (3 February 1468). Gerardy returned to this thesis two years later in amplified form (*Gutenberg Jahrbuch*, 1973, pp. 105–25) but with the date of printing revised to a slightly more possible 1468.

The watermarks in the *Catholicon* do indeed provide a most interesting series of debatable facts. The book was printed on

29

er gloria magnitudo et magnificencia illius et po
teſtas regnū et imperiuȝ in ſecula ſeculoȝ Amen

Altiſſimi preſidio cuius nutu infantium lingue fi
unt diſerte. Qui ȝ nüoſepe puulis reuelat quod
ſapientibus celat. þic liber egregius. catholicon.
dñice incarnacionis annis M ccc lx Alma in ur
be maguntina nacionis inclite germanice. Quam
dei clemencia tam alto ingenii lumine. dono ȝ ȝ :
tuito. ceteris terraȝ nacionibus preferre. illuſtrare
ȝ dignatus eſt ſlon calami. ſtili. aut penne ſuffra
gio. ß mira patronaȝ formaȝ ȝ concordia ȝpor
cione et modulo. impreſſus atȝ confectus eſt.
þinc tibi ſancte pater nato cū flamine ſacro. Laus
et honoȝ dño trino tribuatuȝ et uno Eccleſic lau
de libro hoc catholice plaude Qui laudare piam
ſemper non linque mariam   DEO. GRACIAS

Colophon of the *Catholicon. (Courtesy of the Trustees of The Pierpont Morgan Library.)*

three varieties of paper,[7]—but these sheets were not used indiscriminately, as was usual in the incunabular period. Indeed, the paper used provided three different classes (by quality) of copies.[8] The best paper appears to have been the watermark "Bull's Head"; the middle quality was the Gallizian Seal; the poorest, Crown-Tower marks mixed in definite proportions. This suggests that the printing of the *Catholicon* was not a "hand to mouth" venture, but one that had been carefully planned in advance. This, in turn, implies that the printer needed to have relatively large stocks of paper on hand, since a specific *filigrane* appeared only in the one issue—and appeared throughout that issue. Such a large store of paper, enough to print three issues at one time, runs counter to the beliefs of the filigranologists, who argue that paper, being relatively expensive, had to be bought only as needed.

If we assume the paper to belong to a book printed in 1468, we encounter further difficulties. According to Gerardy (*Gutenberg Jahrbuch,* 1971, p. 23), the paper was made in 1459; thus there must have been a lapse of at least ten years between manufacture and final use.[9] But watermark experts insist that paper was used up in two to three years and that this fact provides the basis for their datings.

A final note on the *Catholicon* watermarks. Gerardy cites the following:

Ich habe die charakteristischen [Gallizian] Zeichen in bisher 32 Druk-
ken erfasst und bin zu dem Schluss gelangt, dass ihre ersten Vorkommen nicht vor 1468 und ihre letzten Vorkommen nicht nach 1479
anzusetzen sind. Allerdings macht [Gerhard] Piccard auf S. 1966[10]
eine Einschränkung: "Es darf angeschlossen werden, dass der Ver-
gleich des Mainzer Catholicon v. J. 1460 mit der ältesten Mentelin-Bibel
Hain 3033 = GW 4203, deren Datierung auf den Zeitraum 1460 bzw.
1461 erkannt ist, ebenfalls die (völlige) Identität der zu beiden Drucken
verwandten Papiere ergab." (*Gutenberg Jahrbuch, 1973*, p. 111)

(Up to the present I have noted the characteristic [Gallizian] marks
in 32 imprints and have come to the conclusion that their occurrence is
not earlier than 1468 nor later than 1479. Nevertheless, [Gerhard]
Piccard in 1966 made this limitation: "It can be concluded that the
comparison of the Mainz Cathlicon of the year 1460 with the oldest
Mentelin Bible, Hain 3033 = GW4203, the dating of which for the
period of 1460 or possibly 1461 is known, gave at the same time the
total identity of the two related paper imprints.) (My translation)

Here Piccard and Gerardy seem to part company, for by their
hypotheses either the Mentelin Bible (with date 1460–1461 in the
Freiburg copy) has to be redated to 1468 or the *Catholicon* firmly
established as 1460, as the colophon proclaims. Thus the 1468 date
of printing, based by Gerardy on other books datable to that year
and containing watermarks identical with those in the *Catholicon*,
is offset by the date of 1460 provided by comparable evidence
discovered by Piccard.

The printed date of 1460 in the *Catholicon* is dismissed by
Gerardy as a mere misprint. Misdatings are, of course, common
in fifteenth-century books.[11] But a "correct" date of: Mcccc lxviii
could never have been fitted into the space available—and this
could not have been the intention of the printer. The date: Mcccc
lx fits comfortably into the room available for it.[12]

A terminal date for the printing is 1465, a date that is found in
the Gotha copy and that represents its acquisition (by purchase or
gift) by the monastery of Altenburg.[13] There seems to be no
reason to doubt the authenticity of this entry—and no one has
come forward to do so.

Afterthought: the reader is invited to draw his own conclusions
as to the value of the evidence provided by watermarks. IF the
evidence of mid-fifteenth-century watermarks is demonstrably so
dubious as seems to be suggested here, is it proper to give
filigranes as great authority as is attributed to them by the experts
now engaged in the dating of the mysterious *Missale speciale?*

Since the above was written, I have received, as the kind gift of
Professor Leonhard Hoffmann, a copy of his article "Ist Guten-

berg der Drucker des Catholicon?" (*Zeitschrift für Bücherfreunde* 93 (1979): 201–13). Hoffmann argues strongly for the view that the *Catholicon* was printed by Gutenberg with the help of his assistant Heinrich Keffer in 1460. He does not consider the possible evidence afforded by the watermarks.

Of even more recent date is the article by Dr. Paul Needham, which appeared not long ago in the *Papers of the Bibliographical Society of America* (vol. 76, no. 4 [1982], pp. 395–456). This study is replete with data on the essays devoted to the *Catholicon* over the past fifty (or more) years. It is recommended that readers familiarize themselves with the fine summaries so that they need not be repeated here.

In seeking a more acceptable solution to the still-unsolved problems concerning the *Catholicon*, Dr. Needham maintains that this work was not printed from movable type but was produced by means of two-line slugs, indissolubly bound together. The slugs, together with those for two slight quartos similarly produced, were then stored to be used as needed. One wonders, however, if this is the *only* solution, and how the printer was able to store and recall individual slugs out of a total of 25,000 or more pieces of metal.

Three points appear from this discussion:

1. If the *Catholicon* was printed by slugs, additional questions are raised. For example, since it was then not printed by movable type, it cannot (by definition) be considered be an incunabulum or be treated as such.

2. If the *Catholicon* or at least the major part of it, was not printed in 1460, how is it to be dated? Is this edition of the *Catholicon* the true Editio Princeps, or not? The answer to this query must be yes and no. The copies with the Bull's Head watermark are judged to belong to the year 1460.

3. The copies with the Galliziani watermark are now assigned to the later 1460s, while those with the Tower-Crown marks are dated as of 1472 or later. This would place the Augsburg edition of 1469 (GW 3183) and probably the two Strassburg editions of circa 1470 (Gesamtkatalog Wiegendrucke 3184 and GW 3185) ahead of most of the copies of the first Mainz printing.

In conclusion, I should like to call attention to two observations made by Dr. Ronald B. McKerrow that may have some bearing on two points brought forward by Dr. Needham.[14] These observations (uneven line endings and the apparently erratic use of special sorts and ligatures) may or may not be significant for the study of the various editions of the *Catholicon*.

Notes

1. For further details, consult Seymour de Ricci, *Catalogue raisonné des premières impressions de Mayence (1445–1467)* (Mainz, 1911).
2. References: GW 3182, Hain 2254, DeRicci 90, Stillwell B19, Goff B-20.
3. Courtesy of the Pierpont Morgan Library. Unfortunately the name of the printer is not given.
4. Margaret Stillwell, *Gutenberg and the Catholicon of 1460* (New York, 1936) (see also her *The Beginning of the World of Books 1450 to 1470* [New York, 1972] p. 17); and Hans Widman, *Der gegenwärtige Stand der Gutenberg-Forschung* (Stuttgart, 1972), pp. 34–41.
5. Theo Gerardy, "Gallizianmarke, Krone und Turm als Wasserzeichen in Grossformatigen Frühdrucken," *Gutenberg Jahrbuch* (1971): 11–23.
6. Widmann's article, "Mainzer Catholicon (GW 3182) und Eltviller Vocabularii: Nochmals zu einer These der Wasserzeichenforschung," in *Gutenberg Jahrbuch*, 1975, pp. 38–48, does not contribute to this discussion.
7. In addition, some ten copies printed on vellum have survived. For the quality of the papers used, see Gerardy, "Wann wurde das Catholicon mit der Schluss-Schrift von 1460 (GW 3182) wirklich gedruckt?" *Gutenberg Jahrbuch* (1973): 105–25.
8. For additional discussion on the papers, see Allan Stevenson, *The Problem of the Missale speciale* (London, 1967), passim.
9. As noted below, Gerardy suggests that the paper was in use until 1479 for a period of some twenty years.
10. This should read: Sp. 319.
11. See my "False Information in the Colophons of Incunabula," *Proceedings of the American Philosophical Society* 114 (1970): 398–406.
12. Alfred W. Pollard's charming English rendering of the colophon reads:

By the help of the Most High, at Whose will the tongues of infants become eloquent, and Who ofttimes reveals to the lowly that which He hides from the wise, this noble book, Catholicon, in the year of the Lord's Incarnation 1460, in the bounteous city of Mainz of the renowned German nation, which the clemency of God has deigned with so lofty a light of genius and free gift to prefer and render illustrious above all other nations of the earth, without help of reed, stilus, or pen, but by the wondrous agreement, proportion, and harmony of punches and types, has been printed and finished.

Hence to Thee, Holy Father, and to the Son, with the Sacred Spirit, Praise and glory be rendered, the threefold Lord and One; For the praise of the Church, O Catholic, applaud this book, Who never ceases to praise the devout Mary.

Thanks be to God.

From *An Essay on Colophons* (Chicago, 1905), pp. 14–15.

13. See Aloys Ruppel, *Johannes Gutenberg; sein Leben und sein Werk* (Berlin, 1947), p. 172.
14. Curt F. Bühler, "The Dictes and Sayings of the Philosophers," Postscript by Ronald B. McKerrow, *The Library*, 4th ser., (1935): 326–29.

2

Musicology and Paper Study—A Survey and Evaluation

Frederick Hudson

At the risk of stating the obvious, it should be made clear at once that the application of paper study to "the Science and Practice of Music" differs little, if at all, from its application to any other discipline.[1] It is as essential for the music scholar to have ready access to the whole literature of papermaking, paper study, and watermarks as for any other researcher and, indeed, my progress would have been painfully slow were it not for the outstanding work of E. J. Labarre and his Hilversum Paper Publications Society series, *Monumenta Chartae Papyraceae Historiam Illustrantia*, supplemented by many specialist writings and collections of a general, nonmusical character. For the music scholar, however, these collections from Briquet onward have one great disadvantage in that their paper studies and watermark illustrations are drawn from State papers, documents, and letters; from manuscript sources; from books and other printed sources; and from maps, engravings, and the like for the larger paper sizes. Rarely, if ever, do they include a manuscript or printed music source. And, as the question of the paper used by most composers has not received intensive investigation, the music scholar is frequently obliged to till virgin soil. Happily, paper is paper no matter by whom it is used, and composers, copyists, music printers, and performers tended to use paper of the larger and more substantial sorts, folded into folios or quartos, the watermarks in which may often be located by *type* in standard collections, even though no identity of outline or parallel period of dating can be established.

At the International Musicological Society's congress held in Salzburg in 1964, Wolfgang Plath stressed that a basis for Mozart scholarship comparable to that established in modern Bach scholarship did not as yet exist.[2] This lead in Bach research goes back over one hundred years to Philipp Spitta's great monograph, *Johann Sebastian Bach*.[3] The exhaustive examination of the original manuscripts conducted by Spitta included the paper and watermarks, which was unique for a music scholar of this period. He presents his watermark evidence in volume 2 under "Anhang A," pp. 767–846 (in the English edition indexed at the end of volume 3 under "Watermarks used by Bach"), and reproduces fourteen watermarks, including three pairs of marks and countermarks, with descriptions of the others. Spitta lists works in which Bach's autograph manuscripts bear the same watermarks; examples are worth quoting as this is a pioneer study in this field:

1. **IMK** with a half-moon countermark in the MSS of forty-one cantatas,
2. shield and crossed swords in three cantatas,
3. **MA** (or **AM**) in thirty-two cantatas,
4. **GAW** (in a cartouche) with a *Post-Horn am Schnur* countermark in ten cantatas,
5. shield flanked each side by a palm tree in eleven cantatas, and
6. **R** / three balls / **S** (in a cartouche) in eight consecutive cantatas for the Church's year from the third Sunday after Easter to Whitsun Tuesday, which, with other evidence, led Spitta to conclude that they were composed the same year.

Bach scholarship received a great impetus by two events in 1950, the bicentenary of the composer's death. The first was the publication of Schmieder's *Bach-Werke-Verzeichnis* (BWV),[4] which supplied a long-felt need and a basis for the second, the promotion of the *J. S. Bach Neue Ausgabe Sämtliche Werke* (NBA). The prospectus promised the issue of five to six volumes per annum (a commitment sadly not adhered to because of a dearth of editors and worsening economic conditions, but nevertheless continuing as a shining example of international collaboration). At the same time that these two events were being planned, the paper historian Dr. Wisso Weiß (following in his father's paths) was making a laborious progress through Bach's original manuscripts, in the several East and West German libraries where they had been dispersed during World War II. Often working in appalling postwar conditions, Weiß noted, traced, collated and catalogued the watermarks he found. The bulky catalogue he completed in the early 1950s and MS copies were deposited in the Bach centers at Göttingen and Leipzig for the use of editors working within the NBA series.[5] Publication is planned within the NBA, but it seems that those responsible are holding this back so that the series can be completed when advantage can be taken of the latest state of research.

Bach research was further accelerated at this time by the founding of the Bach Institute, Göttingen, under the direction of Dr. Alfred Dürr, and its counterpart, the Bach Archives, Leipzig, under Professor Werner Neumann, as well as by parallel work from the staff of the University Music Institute, Tübingen. This activity bore fruit in Alfred Dürr's *Zur Chronologie der Leipziger Vokalwerke J. S. Bachs.*[6] In this work, Dürr provides an annotated catalogue of Bach's enormous output of vocal works based on Wisso Weiß's paper and watermark research and his own examination of notational forms in autograph and other original manuscripts, together with a sifting of all other evidence, and his identification of the handwriting of numerous pupils, friends, and members of Bach's family who wrote out parts and other performing material. The result is the most convincing chronology yet of the period 1723 to 1750, with each work assigned to the Sunday or Feast Day (with date) for which it was composed and/or on which it was performed. Bach research is very much a process of collaboration, and Dürr acknowledges his indebtedness to many other scholars.[7] Added to this is the work of Georg von Dadelsen and his colleagues at Tübingen.[8]

For very many reasons, one of which has to be the gradual recovery from the rigors of the war years, the 1950s saw the launching of projects for the editing and publishing of completely new collected editions of the works of very many composers, for example, the *Neue Mozart-Ausgabe* (NMA), and the *Hallische Händel-Ausgabe* (HHA). Over the decade, this activity extended to the works of nearly all "great" composers and many who had been considered not so great. Possibly because of the enormous progress made in Bach scholarship, music scholars interested in the works of other composers began to apply the same research principles and methods which had achieved such success, including a study of paper. This is certainly true in my case. After working on a volume in the NBA series[9] using Wisso Weiß's catalogue in Leipzig and Göttingen, I found that subsequent work for volumes in the HHA series made it imperative that the paper and watermarks of Handel's autograph MSS, copies in the hands of members of his circle, and the whole range of printed performing material of his lifetime relevant to these volumes be studied and noted, and the findings collated with all other ascertainable evidence. Some aspects of this work are detailed in my essay "The Earliest Paper made by James Whatman the Elder and its Significance in Relation to G. F. Handel and John Walsh," with titles of the HHA volumes and separate critical reports quoted in the footnotes.[10]

At this stage of the survey, tribute must be paid to the outstanding work of the American filigranist, the late Allan H. Stevenson, with special reference to his early study, "Watermarks are Twins," and the influence this had on all workers in this field.[11] He made supremely clear what had not been clear to many, if not most of us: that papermakers at any one vat worked with at least a *pair* of moulds—sometimes three, four or even more—and his list of ten "Points of Difference" which might occur between "twin" watermarks provoked deep thought, sometimes followed by revelation, concerning the nature of "identical" marks, and the uncertainties, even the uselessness as strict evidence of "similar" marks. [See Introduction, pp. 12–13 in the present volume—Ed.] Whether or not Stevenson and other filigranists were the direct cause, the number of music scholars who included paper and watermark evidence in their research, some with greater perception than others, grew steadily during this period. A leader in this field was Professor Jan LaRue of New York, some of whose writings are listed below in chronological order.

Publications of Jan LaRue

"Some National Characteristics of 18th-Century Watermarks." *Journal of the American Musicological Society* 9 (1956): 237. This is an abstract of a paper read to the New England Chapter of the AMS at Wellesley, Massachusetts, on 23 February 1957. He outlines the problems of the authenticity and dating of the MSS of eighteenth-century symphonies, shows that these are helped by study of watermarks, and describes characteristics of Italian, German, and Spanish paper and their watermarks.

"Abbreviated Description of Watermarks." *Fontes Artis Musicae* 4 (1957): 26–28. A working method of recording and describing watermarks, equating brevity with identification.

"British Music Paper, 1770–1820: Some Distinctive Characteristics." *Monthly Musical Record* 87 (1957): 177–80. Describes the
1794 Act of George III (effective to 1811, when it was repealed)
under which papermakers and printers could reclaim export
duty in paper "with a mark commonly called a Water Mark,"
and its usefulness for dating purposes. Offers descriptions of
watermarks but no reproductions.

"Die Datierung von Wasserzeichen im 18.Jahrhundert." In *Mozart-
Kongress-Bericht*, Wien, 1958.

"English Music Papers in the Moravian Archives of North Carolina." *Monthly Musical Record* 89 (1959): 185–88. Among the
thousands of Moravian music copies (mainly of German origin)
are many copies on paper of English origin of the period 1764–
c.1809, distinguished by their white appearance and quality,
and resistance to ink erosion. Watermarks are described but not
reproduced.

"Watermarks and Musicology." *Acta Musicologica* 33 (1961): 120–46.
This is one of the most important and useful paper studies by
LaRue; it is described below in some detail for this reason.

"A Checklist of Watermarks in Dated Musical Sources." *Fontes
Artis Musicae* 9 (1962). This provides descriptions of c. 1000
watermarks but no reproductions.

"Classification of Watermarks for Musicologists." *Fontes Artis Musicae* 13 (1966): 59 ff.

In "Watermarks and Musicology," LaRue makes a valuable contribution which needs more than a passing reference. He begins
by saying that "in the past decade [1950–60] musicologists have
shown a fast-growing interest in watermarks as bibliographical
evidence," and goes on to quote examples from several countries:

Landon, H. C. R. *Symphonies of Joseph Haydn.* London, 1955. *Watermark Master Lists*, pp. 612–14.

Tuttle, Stephen D. "Watermarks in Certain Manuscript Collections of English Keyboard Music." In *Essays in Honor of A. T.
Davison*, pp. 147–58. Cambridge: Harvard University Press,
1957.

Larsen, Jens Peter. *Handel's Messiah, Origins-Composition-Sources.*
London: A. & C. Black, 1957. 336 pp. Gives general descriptions
of watermarks but no reproductions. One of the most valuable
contributions of this book is the identification and codification
of copyists who worked for Handel.

Hartmann, K. G. "Die Handschrift KN 144 der Ratsbücherei zu
Lüneburg." *Die Musikforschung* 13 (1960): 2 ff.

Mendel, Arthur. "Recent Developments in Bach Chronology." *Music Quarterly* 46 (1960): 289 ff.

Bartha, Dénes, and László Somfai. *Haydn als Opernkapellmeister.* Budapest and Mainz, 1960. The *Wasserzeichen-Liste der Papiere in unserem Quellenmaterial*, pp. 435–51, includes 283 dated watermarks with many variants, and three folding tables of full-size reproductions.

LaRue give clear guidance on technical problems in deciphering and recording watermarks, applications of paper evidence, and dating by use of watermarks. He lists national characteristics of watermarks under the headings of ten countries and provides watermark reproductions to illustrate his commentary under each country. Of special interest are his illustrations of deterioration in the wire shapes in the moulds, and therefore in the paper produced in them, which supplements Allan Stevenson's guidance of ten years previously. Finally, LaRue prints full bibliographical details of thirty-four standard works of reference and studies in filigrany, and adds useful comments. It is significant, however, that this important contribution is the sole essay on dating and identification by means of watermarks to appear in the official organ of the International Musicological Society (IMS), *Acta Musicologica*, during the forty-year period 1928 to 1967.

Jan LaRue has also contributed useful articles on watermarks to two leading music dictionaries. The first can be found under *Wasserzeichen* in *Die Musik in Geschichte und Gegenwart* ([MGG] vol. 14, cols. 265–67 [Kassel, 1968]) and includes a pull-out sheet with twenty-five watermark reproductions. The second (with J. S. G. Simmons) under *Watermarks* in *The New Grove Dictionary of Music and Musicians* (vol. 20, pp. 228–31 [London, 1980]) with six outline watermarks and an excellent "Shield with Bend" mark produced by betaradiography. These two articles provide good summaries and could well be consulted as first references for the uninitiated.

To test the significance attached to paper study by music scholars and the use made of it over the period 1950 to 1975, I have examined four research surveys concerned with as many great composers, all published in the official organ of the IMS, *Acta Musicologica*.

Walter Blankenburg's "Zwölf Jahre Bachforschung" covers the period 1953 to 1965.[12] As one might expect, this work describes and documents the almost incredible research activity arising from the bicentenary of Bach's death, the appearance of Schmieder's catalogue (BWV), and the launching of the *Neue*

Bach-Ausgabe. Some twenty-five volumes in the NBA series were published during this period, each with a comprehensive and lengthy *Kritischer Bericht*, published separately, together with pocket scores, vocal scores, and performing parts. Blankenburg lists literally hundred of books and studies concerned with every aspect of the Bach cult and, among the authors who have taken special note of paper and watermark evidence, we read the names of those mentioned above—Dürr, von Dadelsen, Neumann, Mendel and others. We may assume that every editor in the NBA series has examined the paper and watermarks of Bach's original manuscripts relevant to his volume, and consulted one or the other of the MS copies of Wisso Weiß's watermark catalogue deposited at Göttingen and Leipzig (Accredited researchers not working for the NBA are presumably also granted the use of these facilities.) While rejoicing in this enormous activity, one can only regret that Weiß's catalogue remains unpublished and is therefore not available to all interested in this and contemporary fields—especially those who cannot meet the burden of travel and of maintenance during consultation.

Alfred Mann and J. Merrill Knapp's "The Present State of Handel Research" surveys this field from the eighteenth century to 1969.[13] The authors equate the present state of Handel research with that of Mozart, as pointed out by Wolfgang Plath in 1964, namely that neither measures up to that of Bach scholarship. But they add: "One of the principal facts that distinguishes the Handel tradition from the Bach tradition, as well as from the Mozart tradition, is the extraordinary recognition that the composer received both in his lifetime and immediately thereafter." Under the Section "Critical Source Studies," p. 17, the authors state: "Just as the study of the *Messiah* sources [J. P. Larsen, *Handel's Messiah*, 1957] has led to an exploration of the copyists question, the study of the sources for the concertos op. 3 has led to an exploration of the question of watermarks"; and they go on to cite my publications in this field of that time.[14] They end this paragraph with: "As Hudson has pionted out, a work comparable to the catalogue *Papier- und Wasserzeichen der Notenhandschriften von J. S. Bach* is still missing."

Erich Schenk's "Zur Beethovenforschung der letzten zehn Jahre" covers the period 1960 to 1970.[15] Schenk provides an enormous bibliography of literature and editions, to which one of the major contributors is Joseph Schmidt-Görg, formerly director of the Beethovenhaus and Beethoven-Archiv, Bonn, who died April 1981. In 1935 Schmidt-Görg had produced a catalogue of the manuscripts in the Beethovenhaus which included a meager description of the type of watermark in each manuscript. Here Schenk quotes Schmidt-Görg's "Wasserzeichen in Beethovenbriefen"(*Beethoven-Jahrbuch* 5 [1961/1964]: Bonn 1966, 7–74) in which he distinguished over four-hundred watermarks in Beethoven's letters with an introduction and many illustrations. As in his 1935 catalogue, Schmidt-Görg offers no dating conclusions, only classification. Alan Tyson also figures largely in the survey of literature of this decade, but it is in the following period, 1970 to the present time, that he shows himself to be the foremost contributor of paper evidence to Beethoven (and Mozart) source studies. A survey of his outstanding work in this field will be left until later.

D. Kern Holoman's "The Present State of Berlioz Research" surveys Berlioz studies up to 1975.[16] In what is otherwise an exemplary survey there is little specific reference to study of the paper and watermarks of the autograph sources, nor (by default) any mention that such research could have value as contributory evidence. However, in his Ph.D. dissertation, "Autograph Musical Documents of Hector Berlioz, c. 1818–1840" (Princeton, 1974), Holoman *has* taken the paper and watermarks of his sources into account—several thousand hitherto uncatalogued autograph MSS and little-known pages in score, parts, sketches, drafts, etc.—and tabulates the watermarks in Berlioz's scores, with tracings and facsimiles of watermarks. The survey reports that forged MSS, mostly letters, were discovered in 1967–68 (just before 1969, the centennial year of the composer's death, when the forger might expect the highest prices for Berlioziana) and that as many as a hundred forgeries may have been put on the market since that time. A warning about these forgeries was given in a joint letter by Hugh Macdonald, David Cairns, and Alan Tyson to the editor of *The Musical Times* (vol. 110, 1969, p. 32) and, it seems, in this as in all other aspects of research with original sources, paper evidence can be contributory, if not crucial, to conclusions. (We learn that the forger was arrested but that, shortly before this, he may have "hired a second, more skillful forger to produce fake documents of stunningly high quality"!)

From the four surveys above, it seems that Bach research relies largely on the catalogue of Wisso Weiß for paper and watermark evidence and that, in the case of the other three composers, the application of paper study in the periods covered by these surveys *as an avenue of research equal to all others* is the exception rather than the norm.

The following section attempts a survey of the literature concerning music and musicians published from 1961 to 1980, in which paper and watermark evidence is included and discussed, and/or in which watermarks are listed or reproduced. It is not claimed that the coverage is complete but rather that it is representative of a wide range of musical and scholarly interests. As stated above, Alan Tyson's work merits a section on its own.

Wolfe, R. "Parthenia Inviolata: A Seventeenth-Century Folio-form Quarto." *Bulletin of the New York Public Library* 65, no. 6 (June, 1961): 34ff. The only known exemplar of this engraved volume of music for the virginals is held by New York Public Library. It is a companion work to *Parthenia or The Maydenhead of the first musicke that ever was printed for the Virginals* (pieces by Byrd, Bull, and Gibbons), published 1611, with further impressions up to c. 1689. Wolfe's paper and other evidence sheds new light on this *unica*.

Unverricht, Hubert (Köln)." Zur Datierung der Bläsersonaten von Johann D. Zelenka." *Die Musikforschung* 15 (1962): 265–68. This is a study of the paper and watermarks of the MSS of Zelenka's *Six Sonatas for Wind Instruments,* collated with those of signed and dated Zelenka autographs and changing notational forms, leading the author to a compositional period of 1714 to 1723. There are two watermark reproductions, reduced c. 1:2.

Weiß, K. T. *Handbuch der Wasserzeichenkunde* Leipzig, 1962. K. T. Weiß was a pioneer paper historian and the father of Wisso Weiß. This book is an important compendium of information on the science and uses of paper and watermark study—the result of very many years' work in this field. The author died before his work could be published and it was the son who saw it through the press. It seems that Weiß senior was well aware of papermakers' use of at least a pair of moulds at any one vat, and therefore of "twin" watermarks, before Allan H. Stevenson drew attention to this in 1951, but did not gain the credit for this important discovery because of the delay in publication of his book.

Schulte, U. "Besondere katalanische Schöpfformen." *Papiergeschichte* 13 (1963). Useful for all concerned with study of sources arising from this province of Spain.

Mendel, Arthur (Princeton). "Wasserzeichen in den Originalstimmen der Johannes-Passion Johann Sebastian Bachs." *Die Musikforschung* 19 (1966): 291–94. Mendel describes the problems posed by the wealth of original performing parts of Bach's *St. John Passion*, and his division of this material into three "states" representing performances at three different periods for which changes, additions, and deletions were made; his examination of paper and watermarks in Berlin from July 1956 onward helps to differentiate and date the three "states."[17]

Grusnick, Bruno (Lübeck). "Die Dübensammlung. Ein Versuch ihrer chronologischen Ordnung." *Svensk Tidskrift för Musikforskning* 48 (1966): 63–186. This is an investigation of vocal music sources of the Düben Collection in Uppsala University Library (largely copied and gathered by Gustaf Düben, 1624–1690).[18] Evidence for dating consists of watermarks, identification of the different handwritings, and the consecutive numberings Düben added in ink.

Rudén, Jan Olof (Uppsala). "Vattenmärken och musikforskning. Presentation och till ämpning av en dateringsmetod på musikalier i handskrift i Uppsala Universitetsbiblioteks Dübensamling." Ph.D. diss. Uppsala, 1968. 2 vols., mimeographed. This complements Grusnick's work above. The author uses methods and findings by T. Gerady (*Datierung mit Hilfe von Wasserzeichen*, Bückeburg, 1964). Dates for copying are determined for single parts and thin tablature fascicles, but only *terminus ante quem non* could be established for the larger volumes of tablature.

The year 1970 marks the beginning of a new departure and a new understanding on the part of music scholars in the investigation of paper sources, and the application of such evidence to problems of dating in conjunction with all other available evidence. The following studies are examples of this greater perception and application.

Kerman, Joseph (Berkeley), ed. *Ludwig van Beethoven: Autograph Miscellany from circa 1786–1799. British Museum Additional Manuscript 29801, ff.39–162 (The "Kafka-sketchbook").* 2 vols. London: Trustees of the British Museum with the cooperation of the Royal Musical Association, 1970. Volume 1 is a facsimile of the autograph manuscript. On p. xxvi, Table 1 is a Table of Paper-Types (typed by number, paper characteristics of color and size, staff ruling, description of watermarks, and left and right sheets. On p. xxvii, Table 2 gives Folios grouped according to Paper-Types (type, folios, works copied or sketched, with dates). Pp. xxviiiff. provide Notes on Watermarks with references to standard watermark publications. In Volume 2 *(Transcriptions from Autograph)*, there are occasional references to paper and watermarks, especially on pp. 291 to 296.

Idaszak, Danuta (Warsaw). "Znaczenie filigranów dla badán muzykologiocznych nad rekopisami muzycznymi II połowy XVIII wieke" (The importance of watermarks in the examination of music MSS from the second half of the 18th century.) *Muzyka* 15 (1970): 85–90. Three chronological categories emerge for paper originating in Swidnica. Musical works written on this paper by Wojciech Dankowski can be dated (a) from c. 1763, (b) from 1780–86, and (c) from 1792. The MSS contain thirty-eight watermarks.

Clausen, Hans Dieter (Hamburg). *Händels Direktionspartituren ("Handexemplare")*. Hamburger Beiträge zur Musikwissenschaft, Band 7, Karl Dieter Wagner. Hamburg, 1972. 281 pp. This is a descriptive catalogue of the actual conducting scores used by Handel in performance, written out from the autographs by his amanuensis, John Christopher Smith, and other copyists of Handel's circle. After Handel's death, these full-score copies passed through various hands until, in 1868, with Friedrich Chrysander as intermediary, they came into the possession of the Staats- und Universitätsbibliothek, Hamburg. Clausen isolates the various copyists, codifying, and identifying where possible, adding to those codified by Larsen (*Messiah*, 1957), and provides full-size reproductions of some of the watermarks; he generously told me that he received his initial impetus for paper and watermark study from my publications of 1959 to 1963 and that this Hamburg collection offerred him "a very good opportunity" to apply the principles and methods set out in these studies. This is the most important and comprehensive publication yet in relating paper study to Handel sources. It supplies a first paper and watermark catalogue and, when Professor Keiichiro Watanabe (Tokyo), completes his long-term investigation, I hope we may have a complementary catalogue of sources in the Santini Collection at Münster, and of all other autographs and copies arising from Handel's Italian period, 1706 to 1710.

Noblitt, Thomas (Bloomington, Indiana). "Die Datierung der Handschrift Mus. Ms. 3154 der Staatsbibliothek München." *Die Musikforschung* 27 (1974): 36ff. Noblitt's investigation of forty-five different watermarks in the codex revealed that the source was copied c. 1466–c. 1511, and that it was not chronologically ordered. The provenance of the codex, as determined by the watermarks, indicates that many of the forty-two copyists were members of Maximilian's *Hofkapelle* in Innsbruck, and that the source formed part of its repertory.

Wolf, Eugene K., and Jean K. Wolf (Philadelphia, Pennsylvania). "A Newly Identified Complex of Manuscripts from Mannheim." *Journal of the American Musicological Society* 27 (1974): 371–437. The authors describe an extensive search of European and American libraries to find evidence of music used in mid-eighteenth-century performances at the Electoral Court of Mannheim (Johann Stamitz and the Mannheim School), dispersed and thought to have disappeared. This resulted in the location of some 125 MSS which originated in Mannheim. These discoveries are tabulated with titles and contents of MSS, format, paper and watermarks, and copyists. Reproductions of fifteen watermarks are provided and collated, together with reproductions of notational forms, handwriting, etc., and identification of paper sources. This is an important pioneer work in this field. Eugene Wolf has a book on Johann Stamitz due to appear shortly.

Schmidt-Görg, Joseph (Bonn). "Wasserzeichen in Beethoven-Erstausgaben." *Beethoven-Jahrbuch* 9 (1977): 427–52. This is an attempt to identify the watermarks in the early editions of Beethoven's works, but there are no illustrations, no datings attempted, and watermarks are only loosely described.

Brandenburg, Sieghard (Bonn). "Bemerkungen zu Beethovens Op. 96." *Beethoven-Jahrbuch* 9 (1977): 11–25. Opus 96 is the *Violin Sonata in G*. Brandenburg uses Alan Tyson's techniques for analyzing and describing watermarks, and his reproductions of watermarks include drawings and beta-radiographs. This provides a sharp contrast in the approach to and application of paper study in Schmidt-Görg's work above, both in the same periodical issue.

Hudson, Frederick (Newcastle upon Tyne). "The New Bedford Manuscript Part-Books of Handel's Setting of 'L'Allegro.'" *Notes—The Quarterly Journal of the Music Library Association* 33 (March 1977): 531–52. This is an examination of three MS part-books whose owner lives in New Bedford, Massachusetts. Their provenance was traced to Shaw, Lancashire, England, where very many other Handelian MS copies and printed music were located, all belonging to an amateur choral society which flourished between 1740 and 1883. The *L'Allegro* part-books are written on paper made by James Whatman the

Younger (1741–1798), Maidstone, Kent, whose mark "Fleur-de-lys over Shield with Bend over GR" and countermark "J WHAT-MAN" are described and reproduced in beta-radiographs. Using the known facts of Whatman's activities and varying watermarks, investigation of the history of the Shaw Musical Society, its library of MS part-books, their copyists and dates, and other evidence, the date of c 1780 is arrived at for the copying of the New Bedford part-books.

Schmidt-Görg, Joseph. "Die Wasserzeichen in Beethovens Notenpapieren." In *Beiträge zur Beethoven-Bibliographie*, edited by Kurt Dorfmüller, pp. 167–95. Munich, 1978 (published August 1979). This is an updated article similar to that of 1977 (listed above,) though the descriptions now pair mark and countermark. There are still no reproductions or attempted datings, and no distinction between "twin" moulds.

Winter, Robert (USA). "Schubert's Undated Works: a New Chronology." *The Musical Times* 119 (1978): 498–500. Schubert dated some 85 percent of his compositions and drafts, leaving 15 percent undated. Winter pays tribute to Alan Tyson's work on Mozart sources and seeks to apply his methods to Schubert, hoping that study of paper and watermarks may yield valuable clues. In a footnote at the end of this preliminary announcement, he states: "A much more complete study of these and similar problems, with special emphasis on the years 1823–28, can be found in my article in the forthcoming *Schubert Studies: Problems of Style and Chronology*, edited by Eva Badura-Skoda and Peter Branscombe."

Brandenburg, Sieghard. "Ein Skizzenbuch Beethovens aus dem Jahre 1812: Zur Chronologie der Petterschen Skizzenbuches." In *Zu Beethoven*, edited by Harry Goldschmidt, pp. 117–148. Berlin, 1979. An investigation using the same techniques used by Brandenburg in "Bemerkungen zu Beethovens Op 96" mentioned above. Brandenburg includes watermark drawings and discusses them as evidence towards his chronology.[19]

Johnson, Douglas (Philadelphia, Pennsylvania). *Beethoven's Early Sketches in the "Fischhof Miscellany," Berlin Autograph 28*. 2 vols. Ann Arbor: UMI Research Press, 1980. 528 and 292 pp. This work is based on Johnson's doctoral dissertation. It is a close study of the paper on which Beethoven wrote his early (1785–1800) sketches and his autograph scores. As a result of his studies, the author suggests a revised dating for these manuscripts. The chronology is based largely on his paper and watermark investigations, but also takes account of staff-ruling (rastrology), and handwriting. Volume 1, chapter 2, is entitled "Paper Studies: Introduction," chapter 3, "The Vienna Papers: 1792–98," and chapter 4, "The Bonn Papers: 1783–92." In volume 2, pages 264 to 293, the author provides full-size tracings of the watermarks.

————. "Music for Prague and Berlin: Beethoven's Concert Tour of 1796." In *Beethoven, Performers and Critics. The International Beethoven Congress, Detroit 1977,* edited by Robert Winter and Bruce Carr, pp. 24–40. Detroit: Wayne State University Press, 1980. This study attempts to identify those works that Beethoven wrote during his journey to Prague and Berlin in 1796, by distinguishing the papers he obtained while in those cities from the papers he customarily used in Vienna. Johnson's principal criteria are watermarks and staff-ruling.

Publications of Alan Tyson, 1971–1981

Joseph Kerman's pioneering work in his two-volume edition of Beethoven's autograph "Kafka Miscellany" (the so-called *Kakfa-sketchbook*), 1970, has been described above. Shortly after this, Douglas Johnson and Alan Tyson began their collaboration in a systematic investigation of Beethoven's sketchbooks and loose sketchleaves which resulted in the joint essay, drafted in 1971, "Reconstructing Beethoven's Sketchbooks," *Journal of the American Musicological Society* 25 (1972): 137–56. The authors suggest criteria for determining to what extent Beethoven's sketchbooks have been damaged, that is, where leaves have been lost, where leaves have been rearranged, and where extraneous leaves have been added—possibilities which are common to these sketchbooks. The study also shows how surviving, missing leaves can be identified and restored to their original places in the sketchbooks. The principal criteria used are paper and watermarks, staff-ruling, and such additional evidence as inkblots, stitch-holes, and musical continuity. Tyson gave these research principles practical application the next year, in the next work listed below and again, in 1977, in "Das Leonoreskizzenbuch . . ." also listed below.

"A Reconstruction of the Pastoral Symphony Sketchbook (British Museum Add.MS.31766)." In *Beethoven Studies,* edited by Alan Tyson, pp. 67–96. New York, 1973. This takes into account the factor of "twin" watermarks arising from a pair of moulds used alternately at one vat, and their identification. His next study, listed directly below, shows the full use of the techniques for analyzing and describing watermarks which he had developed over the past few years, and includes an appendix in which these principles are codified.

"The Problem of Beethoven's 'First' Leonore Overture." *Journal of the American Musicological Society* 28 (1975): 292–334. This is an expanded version of the fifth lecture Tyson gave at Oxford University, 8 March 1974, where he was James P. R. Lyell Reader in Bibliography from 1973 to 1974. He begins by quoting Beethoven's biographer, Anton Schindler: "Four overtures to one and the same opera! *Fidelio* must surely be a unique case in the long history of opera." He applies his watermark principles and methods to date some of Beethoven's sketches for the overture, *Leonore No. 1*, and claims from the paper evidence that it was written not in 1805 as hitherto thought but at the end of 1806 and in 1807. His appendix, "Ground Rules for the Description of Watermarks," though drafted with special reference to paper types used by Beethoven, could be used as a basic guide to all who seek to add paper study to other avenues of investigation. These "Rules" may be summarized as follows:

1. Describe watermarks in terms of the original *whole* sheet as it left the mould; that is, reassemble the whole sheet from its halves (folio) or its quarters (quarto).

2. When tracing or describing watermarks, always view the watermark from the mould side of the paper, not from the felt-side (I prefer the reverse of this, i.e., as seen when looking down into the mould, but feel that the important point is to make the viewpoint clear and be consistent in description and reproduction throughout).

3. Search for and identify the "twin" forms of paper and watermarks from companion moulds.

4. Number the four quadrants of the sheet from 1 to 4, reading from the bottom left quadrant in a clockwise direction—the "twin" sheet likewise—or, if its watermark is the mirror image, in an anticlockwise direction, starting from the bottom right quadrant.

5. Describe the watermark of an individual leaf by naming its quadrant and its mould, and by referring to the sheet's watermark type.

Tyson's next published studies apply the techniques used for Beethoven research to Mozart sources.

"New Light on Mozart's 'Prussian' Quartets." *The Musical Times* 116 (1975): 126–30; and
" 'La Clemenza di Tito' and its Chronology." *The Musical Times,* 116 (1975): 221–27. These two studies were evidently intended to serve as a guide or introduction to paper and watermark investigation for Mozarteans. The first study suggests that *"Prussian" Quartet No. 1* (and perhaps the first half of *"Prussian" Quartet No. 2*) were not based on drafts from the early 1770s, but were written on paper purchased by Mozart on his way back from Prague in 1789. Tyson shows interconnections between the paper and watermarks of K.589 and K.590, and also offers a way for dating some of Mozart's quartet fragments. The second study connects watermarks with the internal chronology of a work, with illustrations from Mozart and Beethoven, and tries to show in which order Mozart wrote the different numbers of the opera. Paper study suggested that Mozart first wrote the duets and trios, then some choruses and two of the arias for Titus, and finally the solo arias for the other singers. Both studies are intended to throw light on one of Tyson's special interests—Mozart's working methods—what took place in Mozart's "workshop," as it were!

Tyson's next publication concerned the reconstruction of the second sketchbook according to the methods set out in his joint study with Douglas Johnson in 1972.

"Das Leonoreskizzenbuch (Mendelssohn 15): Probleme der Rekonstruktion und der Chronologie." *Beethoven Jahrbuch* 9 (1977): 469–99. Tyson has reconstructed this sketchbook by the same application of research principles as in "A Reconstruction" listed above. A revised version of the original English text of this article is printed in the present volume of essays, pp. 168–90.
"Yet Another 'Leonore' Overture?" *Music & Letters* 58 (1977): 192–203. Hitherto it had been thought that some markings in a copyist's score of the *Leonore No. 2* overture were Beethoven's jottings to turn it into *Leonore No. 3*. From the watermarks, Tyson shows it could not have been copied before about 1810, and that the jottings were probably made in 1814.
"Prolegomena to a Future Edition of Beethoven's Letters." In *Beethoven Studies 2*, edited by Alan Tyson, pp. 1–19. London, 1977. Among the subjects discussed by Tyson are the uses, as well as the limitations, of paper and watermark study in dating Beethoven's letters (pp. 9–12).

A Reconstruction of Nannerl Mozart's Music Book (Notenbuch)."
Music & Letters 60 (1979): 389–400. Here Tyson treats Nannerl's
Notenbuch to the same investigative principles which he applied
to the Beethoven sketchbooks in "A Reconstruction" and "Das
Leonoreskizzenbuch" described above—this is the third reconstruction worked out from the principles Johnson and Tyson
initiated in "Reconstructing Beethoven's sketchbooks." He tries
to determine the original size, the likely concordance in the
book for leaves now in other locations, and the number of
leaves missing.

"The Date of Mozart's Piano Sonata in B flat, KV.^{333}/$_{315c}$: the 'Linz'
Sonata?" In *Musik—Edition—Interpretation: Gedenkschrift Günter
Henle*, edited by Martin Bente, pp. 447–54. Munich, 1980. The
traditional dating for this Sonata has been 1778 or 1779. In his
study of Mozart's handwriting between 1770–1780, Wolfgang
Plath concludes that the autograph was not written in Paris or
even shortly after, but that the handwriting suggests a date of
around 1783–84, "probably not all that long before its first publication in 1784."[20] Tyson believes Plath to be essentially right
and, by the application of watermark and other paper evidence,
fines the date down to "November 1783."

"The Origins of Mozart's 'Hunt' Quartet, K.458." In *Music and
Bibliography: Essays in Honour of Alec Hyatt King*, edited by Oliver
Neighbour, pp. 132–48. London, 1980. Tyson claims that Mozart
began his composition of the "Hunt" Quartet in the summer of
1783, though he did not complete it until November 1784.

"Mozart's 'Haydn' Quartets: The Contribution of Paper Studies."
In *The String Quartets of Haydn, Mozart, and Beethoven: Studies of
the Autograph Manuscripts*, edited by Christoph Wolff, pp. 179–
90. Cambridge, Mass., 1980. This is a study presented by Tyson
at a conference held at Harvard University, 15–17 March 1979. It
is a short survey of the paper and watermarks in Mozart's six
"Haydn" Quartets in which he discusses the relations between
the quartets and the dating of the quartet fragments. He provides tables of the "Paper Distribution in the Autograph" of
these works, and five pages of watermark reproductions.

"The Two Slow Movements of Mozart's Paris Symphony K.297."
The Musical Times 122 (1981): 17–21. Mozart wrote two slow
movements for his "Paris" Symphony, one in 6/8 meter and the
other in 3/4, one replacing the other at a subsequent performance, and it has been generally accepted that the slow movement in 6/8 meter was the original. Tyson adds his paper and
watermark study to all other evidence and suggests that it is the

¾ movement which is the original, and that the ⁶/₈ movement was perhaps finished in Nancy or Strasbourg after Mozart had left Paris.

Two further studies have recently been published, one each on Mozart and Beethoven:

"The Mozart Fragments in the Mozarteum, Salzburg: A Preliminary Study of Their Chronology and Their Significance," *Journal of the American Musicological Society* 34 (1981): 471–510.
"Beethoven's Home-Made Sketchbook of 1807–08," *Beethoven-Jahrbuch* 10 (1983): 185–200. The second study concerns a collection of papers, now broken up and dispersed, on which Beethoven sketched his ideas. Tyson assigns the papers of this sketchbook to the second half of 1807 and the first months of 1808, partly on the evidence of watermarks.

This survey of Alan Tyson's major contribution to paper investigation would not be complete without a report on his fascinating and remarkably successful pioneer presentation on British television of a program which featured a searching study of Mozart's ten "great" quartets, especially from the point of view of how Mozart planned and completed them—this was in the "Discoveries" series, BBC2, London, 16 October 1978. His investigation of the sources took him to the British Library (for the autographs of the quartets), to Salzburg (where the Mozarteum houses most of the fragments), to Vienna, and to a mill in Arnhem where paper is still made by hand—closely followed by the TV cameras and production team. Tyson used beta-radiographs to illustrate the fitting together from scores and fragments of the four quadrants of each sheet, and justified his dating by paper types and watermarks.

As both television and newspapers are not easy of public recall, it is worth quoting from Dr. Stanley Sadie's review of this program in *The Times*, London, 17 October 1978:

> . . . genius at work is a challenge to the investigator; and, though we possess Beethoven's painstaking sketchbooks, for Mozart we have little besides hazy, rose-tinted tales of spontaneous creative combustion.
>
> Mozart did in fact often write quickly, as external evidence shows. But not his mature string quartets. Alan Tyson, using the techniques of paper analysis, has shown that Mozart was not exaggerating when

he wrote, dedicating quartets to Haydn, of a *lunga, e laboriosa fatica*. These six works were slowly pieced together, over two years; batches of paper came and went as Mozart threw off half a dozen piano concertos, a couple of piano sonatas, a symphony, half a mass and much more besides.

The evidence of watermarks, clearly visible through beta-radiography, showed, for example, that Mozart probably changed his mind about the minuet of the "Hunt" quartet, for it is on a leaf foreign to the rest of the work. It showed too that, in an apparent fit of enthusiasm, he bought the paper for his three "Prussian" quartets near a Bohemian paper mill en route from Berlin to Vienna, composed one and a half at once, then stretched the composition of the rest across as many months. Perhaps the special difficulty he experienced in quartet composition stopped him from writing the further three that were planned. That Mozart usually composed on integral batches of paper was demonstrated too, though Dr. Tyson did not explain why the homogenous G minor symphony score was composed, as opposed to just copied, all at once.

This kind of research is especially responsive to the prying eye of the camera, which can see through a watermark, and can stand by and watch while pages are juggled to show their original structure. . . . All this to the sound of Mozart quartets, including some rare fragments; a fascinating programme, presented with uncommon taste.

One of the outstanding Handel scholars of the present time is Keiichiro Watanabe, Professor of Musicology at the Tōhō Gakuen College of Music, and Director of the College Library, Tokyo. For the past ten years, he has been making a study in the greatest possible depth of the paper used by Handel and his copyists in the manuscripts of every work he composed during his Italian period, 1706 to 1710, after he had left Hamburg Opera House, and before he made his first visit to England with a brief stay at Hanover en route. Until recently, the chronology for Handel's Italian period has been hazy—the sequence and lengths of his visits to various cities, how many times he visited them, and dates of composition and performance were not known for sure (though the study of contemporary records and documents by Rudolf Ewerhart[21] and Ursula Kirkendale[22] has helped to fix the dates of some manuscript copies and first performances). Over the period of his research, Professor Watanabe has kindly kept me in touch with his progress and, for the purpose of this survey, has sent me a detailed and well-documented summary of his work, including four analytical lists and approximately one hundred and eighty full-size reproductions of his watermark tracings. He has published two preliminary studies of his work in this field:

Watanabe, Keiichiro (Tokyo). "Die Kopisten der Handschriften von den Werken G. F. Händels in der Santini-Bibliothek, Münster." *Journal of the Japanese Musicological Society* 16, no. 4 (1970): 225–62; and "The Lost Manuscript Copy of Handel's 'Gloria Patri' (The Nanki Music Library, former MS.No.0.52.3). *Tōhō Gakuen College of Music Faculty Bulletin* 3 (June 1977): 42–66.

Watanabe's investigations have taken him to all the centers where Handel's Italian autographs and copies are located—British Library, Fitzwilliam Museum (Cambridge), Royal College of Music, and Royal Academy of Music (London), Bischöfliches Priesterseminar der Universität Münster (Santini Collection, Münster), Staats- und Universitätsbibliothek (Hamburg), Staatsbibliothek Preußischer Kulturbesitz (West Berlin), Library of the Gesellschaft der Musikfreunde (Vienna), Deutsche Staatsbibliothek (East Berlin), and the Koch Collection in the private possession of Dr Georg Floersheim (Basel). His primary sources, the points of departure, are the five autograph MSS in which Handel entered both the dates and places of his completion of composition: *Dixit Dominus* (Rome, April 1707), *Laudate pueri* (Rome, 8 July 1707), *Lungi dal mio del Nume* (Rome, 3 March 1708), *Aci, Galatea e Polifemo* (Naples, 16 June 1708), and *Se tu non lasci amore* (Naples, 12 July 1708).

Watanabe's complicated and painstaking investigation of many thousands of leaves of Handel's manuscripts has resulted in the identification of five different types of paper and their watermarks which, from the evidence of reliable contemporary documents and other evidence, can be accepted to have originated in Italy during the period 1706–11. There are varying identities of outline within these types as follows: (1) *Three Crescents* = 14; (2) *Fleur-de-lys in Double Circle* = 5; (3) *Animal in Circle* = 2; (4) *Fleur-de-lys in a Circle* = 1 only; and (5) *Bird?* = 2. There is also a sixth type without watermarks; concordances in these are established by identity of distances between chain lines and a count of laid lines. The most usual format for these Handel papers is folio (original sheet halved), which, with the "twin" sheet and mark, has meant that Watanabe has searched for and found the four complementary sections making up "twin" sheets. He has tabulated, codified, and arranged his findings from this wealth of papers in five ways: (a) according to paper and watermarks, (b) according to extant dates, (c) according to the participation and identification of copyists, (d) alphabetical arrangement of MSS, and (e) drawings of watermarks—with each of the five sections cross-referenced with sources, titles, etc. He has collated his paper study with all other ascertainable evidence, including (a) Handel's changing handwriting (even during the short period of 1706 to 1710); (b) his use of double hyphens (=) in the verbal text, giving way to single hyphens; (c) the isolation of at least sixteen copyists who worked for him during this period, and the positive naming of some of these; (d) the use of five-nibbed pens by copyists or their hacks, singly or in multiples of three, four, or five coupled five-nibbed pens. His paper research has extended beyond Handel's autographs and copies to the papers of works by other composers, written or copied during the same limited period.

Professor Watanabe has come to some startling conclusions, of which only the following will be mentioned here. The fifteen-movement cantata, *Apollo e Dafne*, has hitherto been considered a typical work (in a whole series of Italian cantatas), composed in Rome at the beginning of Handel's Italian period, when a papal ban on public opera was in force, and the performance of such cantatas in the houses of the wealthy was the alternative. Watanabe has identified five different types of paper in the autograph; one of these types has a "Unicorn" mark, rare in Handel papers, which Watanabe has identified with paper common at this period at the Electoral Court of Hanover. He concludes that the composition of *Apollo e Dafne* marks the point of Handel's departure from Venice early in 1710 and his arrival at Hanover to take up his appointment as Kapellmeister, and that this work was therefore unlikely to have been performed in Italy during Handel's stay there.

We can hope that Professor Watanabe is able to complete his long-term investigation in the near future, and that this major contribution to Handel scholarship will soon be available in published form.

Two further research projects concerned with study of Handel's autograph manuscripts are currently in progress. Donald J. Burrows of Abingdon, Oxfordshire has almost completed his preparation of "A Handlist of the Paper Characteristics of Handel English Autographs"; and Mrs. Martha Ronish, a Fulbright scholar, is working in preparation for a doctoral dissertation within the University of Maryland. Though working independently, both of these researchers are in frequent consultation to their mutual advantage.

As the title of Mr. Burrows' "Handlist" implies, his purpose is to provide a "basic-level aid" to the study of the autograph MSS of works which Handel composed in England from 1711, including, for example, those folios in Italian cantata volumes which are "English" paper. His source locations are the British Library, the Fitzwilliam Museum, and the single volume of cantatas in the Bodleian Library; at present the list covers seventy-five works and groups of works (e.g., the twelve *Chandos Anthems* as one). His preface provides useful information on staff-rulings and various combinations of rastra on paper used by Handel. Though Handel's supplier is unknown, Burrows suggests that this may have been John Walsh, the paper being purchased, already ruled, by John C. Smith for Handel as well as for his circle of copyists. Burrows lists the works chronologically (beginning with *Rinaldo*, first performed February 1711). Under each work he collates the successive gatherings by folio number (or page number for the Fitzwilliam Museum autographs), folio insertions, watermark type, and rastra. For his watermark identification, Burrows adopts the type *sigla* used by Larsen (*Messiah*, 1957, pp. 288–303), and developed by Clausen (*Händels Direktionspartituren*, Hamburg, 1972), with the latter's system of providing measurements of watermarks in millimeters.

Mrs. Ronish has kindly provided the following summary of her research project and its aims:

My work in progress is a catalogue of Handel's autograph manuscripts, of which there are 97 volumes in the British Library, 15 in the Fitzwilliam Museum, and a few fragments in Europe and the United States.

The catalogue will cover such aspects as paper, watermarks, collation, ink and rastral, rubrics (e.g. singers' names, alterations in the music for different performances, corrections and deletions), and other marginalia. It will include an index giving the location of each movement in the autographs, as well as in Chrysander's and the HHA complete works editions. The current lack of such an index is a major handicap for Handel scholars, especially since Handel was notorious for borrowing pieces from one manuscript to use in another; the present-day collation of certain volumes shows a jumble of papers from as many as three decades.

The watermarks in the autographs, coming for the most part from the same papers as are in the conducting scores, will be called by the designations given by Clausen in his *Händels Direktionspartituren*. The numbering of the autographs will follow that of Bernd Baselt's thematic catalogue, *Händel-Handbuch*, (*Thematisch-systematisches Verzeichnis*, HWV, Leipzig, 1978 ff).

My catalogue is expected to be finished by the end of 1983 and, I hope, will be published in co-operation with Donald Burrows and Keiichiro Watanabe.

Mrs. Ronish assures me that she is making a complete set of watermark tracings for each manuscript (using a new light-table aid named after the American musicologist, Rachel Wade). She has completed this task for the Hamburg sources and Dr. Clausen has also made available to her his watermark photographs. She is in touch with Professor Watanabe and, though she will make a personal study of his Italian MS sources on location, she hopes for his collaboration in this field rather than having to duplicate his work. It is Mrs. Ronish's laudable aim to illustrate her dissertation with beta-radiographic reproductions of watermarks, but the extent to which she will be able to achieve this will depend upon the availability of adequate funding.

If the above survey is a true reflection of musical activities subjected to paper study, and of the activities within specific fields by music scholars, then readers with even a minimal knowledge of the literature of music will realize that there are still many fertile areas for future cultivation. This observation is based, obviously, on published writings (though complete coverage of these is not claimed), and on personal knowledge of fields being investigated by colleagues, many over periods of several years, and not as yet published. The known period of paper study by music scholars stretches from the fifteenth to the nineteenth century and includes the works of Buxtehude, Zelenka, Bach, Handel, J. Stamitz and the Mannheim School, Haydn, Mozart, Beethoven, Schubert, Berlioz, and others lesser known. Bach is the sole composer with a complete catalogue of the paper and watermarks in his original manuscripts; others (notably Handel, Haydn, Mozart, Beethoven, and Schubert) are at present represented by intensive investigation of certain areas by a few music scholars. Paper study of Handel sources has been carried out by Hans Dieter Clausen, Keiichiro Watanabe, and a few others, but a change in policy by the Halle Handel Edition authorities has resulted in a decision to cease publication of separate critical reports to the volumes in this collected edition. This has had the unfortunate consequence that the work of several editors who have made an intensive study over the past ten years of the paper and watermarks of Handel's autograph MSS, firsthand copies, and original prints remains unpublished in the HHA—Professors Howard Serwer (University of Maryland), J. Merrill Knapp (Princeton), Alfred Mann (Rochester, N.Y.), and I among these unfortunate editors! Obviously the ideal would be a paper and watermark catalogue of the works of *every* composer or, at least, of the very many composers whose works are receiving renewed practical and critical attention in the post-1950 collected editions. For each composer this would be a major undertaking on the part of a paper historian or bibliographer (music scholars are frequently musicians as well and are unwilling to devote several years solely to bibliographical tasks!), but, if the will and the means were there, a commissioned paper and watermark catalogue for each collected edition would seem to be as reasonable as it is necessary. Meanwhile, music scholars with the greatest perception will continue to combine paper study with all other aspects of investigation in their chosen fields.

Beta-radiographic facilities are now available in many libraries in Britain, Holland, Sweden, Denmark, Finland, other western European countries, and North America. It seems, however, that in the Federal Republic of Germany, such facilities are used more by scientific institutions and departments than by libraries, the decision whether or not to purchase a beta-radiographic source depending frequently on the attitude of the director or librarian to the use of "radio-active material." One may hope that the use of this simple and effective method may soon be widely accepted (especially with the use of such a harmless isotope as Carbon 14) and proliferate to the benefit of scholarship in this country. As I approach the end of this survey, confirmation has reached me that, with the cooperation of the Zentrales Isotopenlabor department of the University of Göttingen, the Bach Institute, Göttingen, now has the benefit of beta-radiographic facilities for the reproduction of paper and watermarks, and that the hand tracings of watermarks in Wisso Weiß's catalogue, *Papier und Wasserzeichen der Notenhandschriften von Johann Sebastian Bach*, have now been replaced by beta-radiographic reproductions. This is indeed welcome news for Bach scholarship and a major cause of congratulation to the Bach Institute and its director, Dr Alfred Dürr. At the time of preparing this essay for the press I learn that this historic catalogue is scheduled for publication in two volumes in 1985.[23]

Notes

1. Phrase is taken from the title of John Hawkins' *A General History of the Science and Practice of Music*, 5 vols. (London, 1776; reprint, 3 vols., Novello, London, 1853 and 1875).

2. *Bericht über den Internationalen Kongress Salzburg 1964*, vol. 1, p. 47 ff.

3. Philipp Spitta, *Johann Sebastian Bach* (Leipzig: Breitkopf & Härtel, vol. 1, 1873, vol. 2, 1880; English ed. in 3 vols., trans. Clara Bell and J. A. Fuller-Maitland (London: Novello, 1899).

4. Wolfgang Schmieder, *Thematisch- Systematisches Verzeichnis der Musikalischen Werke von Johann Sebastian Bach* (BWV) (Leipzig: Breitkopf & Härtel, 1950), 747pp.

5. Wisso Weiß, *Papier und Wasserzeichen der Notenhandschriften von Johann Sebastian Bach*, MS copies deposited at Bach-Institut, Göttingen, and Bach-Archiv, Leipzig.

6. Alfred Dürr, "Zur Chronologie der Leipziger Vokalwerke J. S. Bachs," in: *Bach-Jahrbuch*, 44 (1957):5–162; (also published separately, Kassel, Basel, Tours, London: Bärenreiter, 1958; 2d ed.: *Mit Anmerkungen und Nachträgen versehener Nachdruck aus Bach-Jahrbuch 1957*, Bärenreiter, 1976).

7. Peter Wackernagel (Berliner Stimmen-Signaturen), Werner Neumann (Leipzig), Wolfgang Plath (Tübingen), and Arthur Mendel (Princeton).

8. Georg von Dadelsen, *Bemerkungen zur Handschrift Johann Sebastian Bachs, seiner Familie und seines Kreises* (Trossingen: Hohner Verlag, 1957). This is Heft 1 of the *Tübinger Bach-Studien*, with further studies by Walter Gerstenberg, Paul Kast, and Wolfgang Plath, under Gerstenberg's general editorship. There is also von Dadelsen's Habilitationsschrift, *Beiträge zur Chronologie der Werke Johann Sebastian Bachs* (Tübingen, 1958).

9. *Neue Bach-Ausgabe*, Band I/33 *(Trauungskantaten)*, ed. Frederick Hudson (Kassel, Basel, London, New York: Bärenreiter, and Leipzig: VEB Deutscher Verlag für Musik, 1957), with separate *Kritischer Bericht* (Kassel and Leipzig: Bärenreiter and VEB Deutscher Verlag für Musik, 1958).

10. *The Music Review,* 38 (1977): 15–32 and 7 plates.

11. Allan H. Stevenson, "Watermarks are Twins," *Studies in Bibliography* 4 (1951–52): 57–91.

12. Walter Blankenburg (Schlüchtern), "Zwölf Jahre Bachforschung," *Acta Musicologica* 37 (1965): 95–158.

13. Alfred Mann (New Brunswick, N.J.) in collaboration with J. Merrill Knapp (Princeton, N.J.), "The Present State of Handel Research," *Acta Musicologica* 41 (1969): 4–26.

14. Frederick Hudson (Newcastle upon Tyne), "Concerning the Watermarks in the Manuscripts and Early Prints of G. F. Handel," *Music Review* 20 (1959): 7–27; and "Wasserzeichen in Händelschen Manuskripten und Drucken (Wasserzeichen in Verbindung mit anderem Beweismaterial zur Datierung der MSS und frühen Drucke G. F. Händels)," *Händel-Konferenzbericht 1959* (Leipzig, 1961), pp. 193–206. In the former the nineteen watermark reproductions are full-size; in the latter and in the *Kritischer Bericht* to Opus 3 (Kassel and Leipzig, 1963), these are reduced 1:2.

15. Erich Schenk (Wien), "Zur Beethovenforschung der letzen zehn Jahre," *Acta Musicologica* 42 (1970): 83–109.

16. D. Kern Holoman (Davis, California), "The Present State of Berlioz Research," *Acta Musicologica* 47 (1975): 31–67.

17. After twenty years' research, Arthur Mendel's edition of the *Johannes-Passion* (BWV 245) was published in 1973 (NBA Serie II/4, full-score xiii + 269pp, with detached facsimile of first eleven pages of Bach's fair-copy score, *P 28,* c. 1739); reviewed in *Music & Letters* 56 (1975): 91–95.

18. The Düben Collection, Uppsala, includes nine works by Franz Tunder (1614–1667), and everything that survives of the vast cantata output of Dietrich

Buxtehude (c. 1637–1707), his successor at Lübeck's *Marienkirche*—the sources in Lübeck, Berlin, Wolfenbüttel, and Brussels merely duplicate copies of certain works. The background to this collection and the many composers represented in it are described in my essay, "Franz Tunder, the North-Elbe Music School and its Influence on J. S. Bach," *The Organ Yearbook* 8 (1977): 20–40.

19. During the Handel Bi-Centennial Festival, Halle, 11–19 April 1959, Professor Harry Goldschmidt and I had the opportunity of discussing techniques of paper study and their application to research problems, and of driving down to Greiz to consult with Dr. Wisso Weiß. Professor Goldschmidt declared his hope that watermark evidence would help to solve problems with Schubert sources, but so far does not seem to have published the results of such work with this composer or with Beethoven.

20. Wolfgang Plath, "Beiträge zur Mozart-Autographie II. Schriftchronologie 1770–1780," *Mozart-Jahrbuch 1976/77*, 1978, pp. 131–73, especially p. 171.

21. Rudolf Ewerhart, "Die Händel-Handschrift der Santini Bibliothek in Münster," *Händel Jahrbuch 1960*, pp. 111–50.

22. Ursula Kirkendale, "The Ruspoli Documents on Handel," *Journal of the American Musicological Society* 20 (1967): 22–73.

23. I express my gratitude and warmest thanks to friends and colleagues who have helped with information for this survey—to Dr Ruth Blume (MGG Schriftleitung, Kassel) for a report on the current state of beta-radiography in the Federal Republic, to Professor Keiichiro Watanabe (Omiya and Tokyo) for the trouble he has taken to inform me of the precise state of his research, to Professor Douglas Johnson (Rutgers University, New Jersey) for writing to me about his work, and to Dr. Alan Tyson (All Souls College, Oxford) for generous help with information and comments on his own work and that of certain other music scholars who are also filigranists.

Bibliography of Works Dealing with Beta-Radiography of Watermarks, 1960–1972

Erastov, D. P. "Beta-radiografičeskij metod vosproizvedenija filigranei i dokumentov." In *Novije metodij resttavrasii knig: sbornik rabot zu 1958 god*, pp. 139–48. Moscow and Leningrad, 1960.

Grönvik, Anna. "The Paper Museum at the Finnish Pulp and Paper Research Institute." *Papper och trä* (1966): 519–24.

Liljedahl, Gösta. "Om vattenmärken i papper och vattenmärkesforskning." *Biblis* (1970): 91–129, special references, 116–17).

Milveden, Ingmar. "Neue Funde zur Brynolphus-Kritik." *Svensk tidskrift för musikforskning* 54 (1972): 5–51, special reference, 19.

Nordstrand, Ove K. "Beta-Radiographie von Wasserzeichen." *Papier Geschichte* (Mainz) 17 (June 1967): 25–28.

Philippi, Maja. "Wasserzeichen auf der Spur. Radioisotope im Dienste der Geschichtswissenschaft. Wertvolle Leistung einer Kronstädter Forschergruppe." *Neuer Wege* (Bucharest) 18, no. 581 (15 December 1966): 1.

Putnam, J. L. *Isotopes,* pl. 15. London: Penguin Books, 1960.

Simmons, J. S. G. "The Leningrad Method of Watermark Reproduction." Review of Erastov's article. *The Book Collector* (London) 10 (1961): 329–30.

———. "The Delft Method of Watermark Reproduction." *The Book Collector* 18 (1969): 514–15.

Stevenson, Allan. *The Quincentennial of Netherlandish Blockbooks.* London: British Museum, October 1966. (8 pp., 4 radiographs, first use of beta-radiographs in a journal as evidence of date).

———. "Beta-Radiography and Paper Research," pp. 159–68. Paper delivered at the Seventh International Congress of Paper Historians. *Communications,* Oxford, 1967.

———. *The Problem of the Missale Speciale,* special references, pp. xi, 44, 66–68, 102, 124, 172, 198. London: Bibliographical Society, 1967.

Tydemen, P. A. *British Paper and Board Industry Research Association Bulletin 36,* October 1964, pp. 10–21. Kenley, Surrey.

———. "A simple method for contact Beta-radiography of paper." *The Paper Maker & British Paper Trade Journal* (Kenley, Surrey) 153 (June 1967): 42–48, 65.

van Ooij, W. J. "Betaradiografie van papier met behulp van vlakke betabronnen." *Papierwereld* 24, no. 3 (1969): 67–73.

3

Early Dated Watermarks in English Papers: A Cautionary Note

Hilton Kelliher

To those who have cause to study English books, pamphlets, and written records of the late eighteenth and early nineteenth centuries, the most important of several acts framed in the reign of George III to regulate the duties payable on paper is one of 4 April 1794 that secured the incorporation into papermakers' moulds of dated watermarks.[1] This Act stipulated that in order to be eligible for the drawbacks allowed on the excise duties paid on newly manufactured papers that were subsequently exported, all such papers must have

> visible in the substance thereof a mark commonly called a *Water Mark,* of a date of the present Year of our Lord in the following figures, 1794, or in a like manner of some subsequent Year of our Lord.[2]

This provision held good whether the paper exported was in loose quires or in book form. As the prime purpose of the government was to ensure that drawbacks should apply only to papers manufactured after the Act became law, it did not require in so many words that the date in each stock be henceforth that of the year in which it was actually made, though this was probably the intention. Consequently, papermakers exercised their own discretion about altering the dates in subsequent years, many retaining the "1794" mark in their moulds until at least the turn of the century. Besides recording *en passant* the existence of such a stray mark as "James Whatman Turkey Mill 1792," Thomas Balston, writing of those employed after the Act by the firm of Hollingsworth and

Balston, noted that "except for two instances of 1797 I have found, in a fairly wide search, no date between 1794 and 1801."[3]

Despite this sort of irregularity, bibliographers and others have had ample cause to be grateful for the advent of the dated mark, since, treated with due caution, it may considerably facilitate the dating of otherwise unattributed books, printed ephemera, letters, and documents for a period extending many years after its widespread (though not invariable) adoption by law. Experience has established useful, if by no means infallible, guidelines in this matter. The late Edward Heawood, after examining systematically some eighty papers belonging to the early nineteenth century, concluded that the average interval between the manufacture of papers and their use was "not quite three years."[4] This is borne out by an investigation, undertaken for present purposes, of some writing papers preserved in the British Library,[5] from which it emerges that in the comparatively few instances where the date in the paper corresponds with the year of its use, any given document is unlikely to have been written before very late June.[6] A possible reason for this fairly consistent time lag may perhaps be sought in the retail processes by which papers were stored in warehouses by manufacturers or wholesale stationers until needed for sale to customers. Although the date in the paper is, therefore, not normally expected to coincide exactly with that of writing or printing, it has generally been felt safe to assume that these dated marks provide at the least a reliable *terminus a quo* for the works in which they appear: in other words, that paper bearing the date of a particular year could not have been manufactured, much less be available for use, during the previous one. A single and apparently well-authenticated instance has, however, come to light in which this assumption is shown to be unwarranted. As the implications of this chance discovery stretch far beyond the individual case, the facts are stated here in the hope that filigranists may be able to add something to the elucidation of this matter, and perhaps even to reveal other instances from sources originating at this or at any subsequent period.

In 1806 the Sheffield journalist and poet James Montgomery (1771–1854) issued, as his second book of verse, a collection entitled *The Wanderer of Switzerland, and Other Poems*. Richard Garnett, writing in the *Dictionary of National Biography*, remarked that the title poem,

> founded upon the French conquest of Switzerland, took the public ear at once, probably on account of the subject, and from the merit of

some of the miscellaneous pieces accompanying it, especially the really fine and still popular lyric "The Grave". The principal poem is as a whole very feeble, though a happy thought or vigorous expression may be found here and there. The volume nevertheless speedily went through three editions, and its sale was not materially checked by a caustic review from the pen of Jeffrey (*Edinb. Rev.* January 1807) which indeed gained Montgomery many friends.[7]

The first edition, printed at Montgomery's own office for the publishing firms of Vernor, Hood and Sharpe, and Longman, Hurst, Rees and Orme, was advertised in the *British Press* of 9 February 1806. By the summer, the publishing rights had been transferred exclusively to Longman, whose records, now held in the Library of the University of Reading,[8] show that in July he brought out a second edition totaling 1000 copies, printed jointly by George Cooke of St Dunstan's Hill and the relatively young firm of Wood and Innes, of Poppins Court, Fleet Street.[9] This volume was announced in the *Monthly Literary Advertiser* for 9 August. So great was the demand for the work that a third edition, numbering 2000 foolscap octavo copies retailing at five shillings each in boards, was set up three months later by Wood and Innes and by Caleb Stower of Paternoster Row, the printing being accomplished at some time between 13 October, the date of the author's preface, and 4 November, when advertisements appeared in the *Sun* and the *British Press*.[10] The employment of two printers, though not uncommon in a virtual reprint such as this, argues a need for some urgency in the matter; and the dating is corroborated by the printing account that stands in Longman's Impression Book No. 3, opening 26, though the arithmetic is not entirely accurate.

Montgomerys Poems—3ᵈ Edit—2000—Nov 1806

Printing	4½ Sheets—@ 4. 14. 0 — Wood		21	3	—
	Corrections Postage — Labels &c		1	12	—
	3 Sheets Stowers @ 5 - 2- 0		14	2	—
Paper	30 Rm Copy 38/6	Longman & D	57	15	—
	Hotpressing		3	15	—
			£98	7	—

Despite the appearance of a similar calculation under the date of 12 January 1807 in Longman's Joint Commission and Divide Ledger for 1803–1807 (opening 153), there is no evidence of any second impression being set up at this later period; and, though possible, it is unlikely that the 2000 copies printed in October would have been sold out in a mere two months. After the third edition, demand for the work seems to have leveled off, and in June and December 1807, further sums are recorded as being paid for advertising.[11] Between July 1808 and 1826, however, another seven editions were published by or for Longman, to say nothing of thirteen separate American issues that appeared between 1807 and 1817.[12]

The paper that was used for the printing of the third edition of *The Wanderer of Switzerland* is a wove one of good quality, and though it was used in the autumn of 1806, it bears the dated countermark "H S / 1807." It is as usual possible to discern two distinct marks of the same general type emanating from the twin moulds that were used in production of this stock of paper. One of these, found, for example, on sig. G, pp. 129, 130, of the British Library copy (pressmark 11659.a.38) of Montgomery's *Wanderer*, is a regular and quite sharply defined mark: the other, occurring, among other places, on sig. D, pp. 57 and 58, shows blurring of the letters and apparent damage to the figure "8."[13] The badly aligned final "7" in the latter may perhaps imply that it replaced a figure belonging to an earlier mark such as "H S / 1806," and this could explain what appears to be rather excessive wear and tear at this comparatively early stage in its life. These twin yet quite distinct countermarks also occur in gatherings standing side by side in other works, listed below, that were printed on this stock.

As the above account shows, the price of the stock was thirty-eight shillings and sixpence per ream, and the suppliers were the firm of Longman and Dickinson, who traded as stationers at 39 Ludgate Street. George Longman was the publisher's uncle, and John Dickinson a young man setting out on the career that led to papermaking and the founding of the famous firm that bore his name.[14] Unfortunately, until cataloguing of the firm's archives is complete, it will not be possible to trace further details, if any survive, of this stock.[15] However, Thomas Norton Longman must have bought altogether at least two hundred reams, for he distributed it in turn to various printers for use in William Wordsworth's *Poems, in Two Volumes* (printed by Wood and Innes, January–April 1807), Martin Master's *Progress of Love* (Hazard, February), the anonymous *Pleasures of Human Life* (McCreevy, March), Porter's *Sidney's Aphorisms* (Stower, April) and Boyd's *Petrarch's Triumphs* (Wood, April–May 1807). Whether he bought this paper in one job lot or over a period of time is not known, though possibly its last appearance in a work executed for him is

Watermark from Montgomery's *Wanderer,* pressmark 11659.a.38, pp. 129, 130. *(Courtesy of The British Library.)*

Watermark from Montgomery's *Wanderer,* pressmark 11659.a.38, pp. 57, 58. *(Courtesy of The British Library.)*

in Edmund Cartwright's *Letters and Sonnets* (Pople, October 1807). The occurrence in copies of Miss S. Evance's *Poems*, also printed for Longman by Mc'Creevy in December 1808, of the dated mark "H Salmon / 1807" may encourage us to identify the manufacturer of the earlier stock as Henry Salmon of Langcliffe Mill near Settle, in Yorkshire.[16]

Only four copies, all of them bearing the "H S / 1807" watermark, of the third edition of Montgomery's *Wanderer* have been examined for present purposes;[17] yet, although in bibliographical terms the sample is very small, the known facts of the case strongly support a single printing in October 1806 and lead one to believe that these cannot simply be strays from some other impression. It is admittedly striking that the other known examples of this stock so far traced among Longman's publications should crop up only between January and May 1807, beginning with Wordsworth's *Poems, in Two Volumes*, the copy for which was received in batches at Poppins Court and printed off sheet by sheet from early January to April, immediately after correction of the individual proofs by the author.[18] Most probably the manufacturer's or publisher's warehouseman, having accidentally released a batch of this stock in October 1806, held back the remainder until the following year in accordance with a custom that presumably obtained in the printing and papermaking trades at this time: after all, it would scarcely have been feasible to apply for a drawback, if need arose, on paper that was postdated. What we have here, then, must be very much an exception to established practice. All the same, it is necessary to attempt an explanation of the seeming anomaly by which paper dated 1807 could have been in existence by October 1806. Perhaps the simplest explanation depends upon a purely practical consideration. Given the loose terms of the 1794 Act, papermakers were under no obligation to alter their marks with the change of year, but rather as those marks, or the moulds to which they were attached, began to show serious signs of deterioration. Moreover, when the time came to make the change, they were not restricted—other than by the regulation governing drawbacks—to the date of the current year, but could move backward or forward in the calendar as it suited them. Thus the manufacturer of the paper that was used for *The Wanderer of Switzerland*, when replacing his moulds or marks in the late summer or early autumn of 1806, was quite at liberty to incorporate the date of the following year. He may have done this because he was intending to lay up some stocks in reserve for the future, or because he expected the moulds to last well into 1807. Even though it has been estimated that the average life of a pair of moulds at this period was six months,[19] which means that they would normally have become obsolete by April 1807, paper produced in them between October and December would still have been current if sold in the following year. If, in addition, this particular manufacturer did not make a practice of exporting personally, there was no reason why he should not follow his own inclination in the matter.

The system of tax concessions then prevailing must account for the apparent rarity of postdated papers. No other instances of this phenomenon seem to have been recorded by students of paper-making, and an examination of upward of one hundred titles from those that were published by Longman between 1794 and 1809—taking, that is, books that were printed in the last three months of each year—has failed to throw up any similar example. Yet although there is no reason to suppose that postdating will ever be a frequent factor in the dating of books and documents, the possibility of its occurrence will have to be borne in mind in the future, whenever watermarks are quoted as evidence.

Notes

1. Statute of 34 Geo. III, cap. 20. See further, Rupert C. Jarvis, "The Paper-Makers and the Excise in the Eighteenth Century," *Library*, 5th ser., 14 (1959): 100–116 (esp. 114).

2. Statute of 34 Geo. III, cap. 20, sec. 31.

3. Thomas Balston, *William Balston, Paper Maker* (London, 1954 [1955]), pp. 16–17n. 1; and see pp. 164, 165.

4. Edward Heawood, *Watermarks, Mainly of the 17th and 18th Centuries* (Hilversum, 1950), p. 31.

5. The following collections of letters and papers, preserved in the Department of Manuscripts, have been systematically examined: Cumberland Papers, 1795–1820 (Additional MSS 36498–36507); Liverpool Papers, 1794–1822 (Add. MSS 38472–38474); Huskisson Papers, 1794–1821 (Add. MSS 38734–38742); and Grenville Papers, 1794–1819 (Add. MSS 41856–41858 and 59367–59404).

6. The sole exception found was a letter of 28 May 1819 in Add. MS 36507, f.114, which bears a Whatman counter mark dated to that year. This instance occurs a quarter of a century after the 1794 Act, by which time practices may have changed.

7. Edited by Sidney Lee (London, 1894), 38:318.

8. Thanks are owed to the Longman Group, the owners of these papers, for permission to consult and to quote from them, and to the staff of the Archives Department at Reading University Library for their cheerful assistance.

9. By 6 Nov. 1805, when Charles Wood registered his press, his partner John Innes was only six months out of his apprenticeship. Together they worked for Longman in 1806 on small reprints or new editions of four plays (May and

December), on the third and fourth volumes of *The Mysteries of Udolpho* (November) and on Andrew Tooke's *Pantheon* (by 13 November). Their names last figure in his Impression Books in November 1808. See William B. Todd, *A Directory of Printers . . . London and Vicinity, 1800–1840* (London, 1972), pp. 104, 215; D. F. McKenzie, *Stationers' Company Apprentices, 1701–1800* (Oxford, 1978), p. 155, no. 3594; and Longman's Impression Book No. 3, openings 20, 21, 31, 45, 48, 63, 148, 152 and 154. A useful reference work is Alison Ingram's *Index to the archives of the House of Longman, 1794–1914*, Cambridge: Chadwyck-Healey, 1981.

10. Under the date of 9 October 1806, Longman's Joint Commission and Divide Ledger 1803–1807, opening 153, mentions payment of two shillings and threepence for "Carriage of Parcel," that is, most probably, the sending of the author's revised copy from and to London for revision and printing.

11. Joint Commission and Divide Ledger, 1803–1807, opening 153. See also Divide Ledger 1807–1828, opening 44.

12. British Library, *General Catalogue of Printed Books to 1975* (London, 1979–), and *The National Union Catalog: pre-1956 Imprints* (Mansell, 1975), 392:311, 312.

13. Beta-radiographs of these marks, reproduced by kind permission of the British Library Board, will also be found in *The Manuscript of William Wordsworth's 'Poems, in Two Volumes' (1807): a facsimile, with an introduction by W. H. Kelliher* (London: The British Library, 1984), p. 47.

14. Details of his career to 1809 are given in Joan Evans' *The Endless Web* (London, 1955), chap. 1, pp. 1–4.

15. I am grateful to messrs. R. F. Davis and D. Lole of DRG, the modern representative of John Dickinson & Co. Ltd., for information concerning the Dickinson archives held at Apsley Mill, Hemel Hempstead.

16. Alfred H. Shorter, *Paper Mills and Paper Makers in England 1495–1800* (Hilversum, 1957), p. 250. Another possible candidate is Henry Skeats of Romsey Mill, Hampshire: see Shorter, *Paper Mills*, pp. 172, 173.

17. The copy in the British Library Department of Printed Books has the pressmark 11659.a.38. Others are in the possession of the present writer and of Professor Mark Reed of Chapel Hill, North Carolina, while a copy bearing the bookplate of Pauncefoot Duncombe (1804–49) of Great Brickhill was offered for sale by the London bookseller Arthur Page in autumn 1982.

18. See Kelliher, *op cit.* (above n.13), pp. 64, 65, and *The Manuscript of William Wordsworth's 'Poems,' 'Poems in Two Volumes,' and Other Poems, 1800–1807 by William Wordsworth*, ed. Jared Curtis (Ithaca: Cornell University Press, 1983, pp. 716, 717.

19. Thomas Balston, *James Whatman, Father & Son* (London, 1957), pp. 60, 120: quoted in Philip Gaskell's *A New Introduction to Bibliography* (Oxford, 1972), p. 63 n. 12.

4

The Reina Codex Revisited

John Nádas

Since its sale by Signor Reina in Milan on 15 December 1834, the manuscript Paris, Bibliothèque Nationale, nouv. acq. fr. 6771 (Reina Codex) has assumed an important position at the heart of our understanding of late-fourteenth- and early-fifteenth-century Italian and French secular polyphony. It has been the focus of investigation in numerous studies and editions, in which scholars have delved into questions concerning the physical aspects of the source, its scribes, repertory, and the character of its musical text.[1] In addition, all of the music in the manuscript has been made available in modern transcription.[2] Although much can be agreed upon, these studies and editions have left us with conclusions that vary considerably; most important, they present solutions to problems of structure and repertory that remain controversial and that raise new questions not only about the makeup of the manuscript, but also about the nature of the readings it transmits.

This essay focuses on the results of a fresh examination of the physical features of this manuscript: the identification of watermarks (a point that has not previously been properly understood); characteristics of ruling; the binding; and, finally, the distinguishing of scribal hands.[3] The results, differing significantly from solutions published thus far, bring us closer to an appreciation of the working habits of the scribes and the nature of their collaboration, and hence closer to a critical reading of the musical texts. Against this backdrop of observations, I shall make some points concerning scribal contributions to the variants preserved in the collection.[4]

Pioneering work on the Reina Codex was undertaken by Johannes Wolf in his monumental *Geschichte der Mensural-Notation von 1250–1460*,[5] in which he focused primarily on a description of the contents and its notation. Friedrich Ludwig, both in a review of Wolf's survey and also in the introduction to his own edition of the works of Guillaume de Machaut, corrected errors in Wolf's description of contents and added his own analysis of the makeup and scribes of the Reina Codex:[6] (1) the first seven gatherings (fols. 1–84), with a space reserved for writing measuring 20 × 17.3 cm, consist of an Italian and French fourteenth-century repertory rich in *unica;* (2) a second section was compiled later in the Quattrocento, and consists of French works of the early fifteenth century. Ludwig's description was subsequently refined by Leo Schrade, who distinguished three sections, each characterized by repertory and scribes: "the Italian Trecento portion, the simultaneously collected French repertory of the 14th century, and the fifteenth-century supplement."[7]

Today, much of the controversy surrounding the Reina Codex is embodied in the two full-length studies of the source by Kurt von Fischer and Nigel Wilkins.[8] To aid the discussion to follow, I shall here outline their respective arguments.

Von Fischer's most valuable contribution remains his 1957 description, in which the manuscript's importance and general place of origin were asserted. In all fairness to von Fischer it must be stated at the outset that in matters codicological his study reflects the state of the art in the 1950s; it represented the field of music history, which unfortunately lagged—and in some ways continues to lag—behind sister disciplines in the study of its source materials. In the preface to his inventory,[9] von Fischer noted the following:

> 1. Part 1 of the codex (gatherings 1 to 5, consisting primarily of 14th-century works in Italian notation with Italian texts) was copied c. 1400 by hands A (fols. 1–39v, 43–44, 47v–52v), B (fols. 39v–41, 45v–46), and C (fols. 44v–45). In some cases it was difficult to distinguish among the different hands (e.g., fols. 9v–10).
>
> 2. Part 2 (gatherings 6 to 7, consisting almost entirely of fourteenth-century works in French notation with French texts) was the responsibility of two new scribes, writing c. 1400: D (fols. 53–62v, and additions on 12v–13, 46v–47, 65v–66, 72v–73, and 77v, as well as texts on folios 65 and 70 [D also compiled the index on fols. 126v–127]) and E (fols. 63–84v, and possibly the instrumental arrangements on inserted folios 85r and 85v [noting that the scribe was quite possibly Italian, judging from corruptions in the texts]).[10]

3. Part 3 is a supplement, added c. 1430–1440 by Hand F (fols. 89v–119). Von Fischer considered this section of the manuscript as consisting of two gatherings (8 and 9) and various inserted folios.[11]

4. An original verso-side foliation (i.e., one that counts openings in the collection) in roman numerals extends from the beginning of the source to folio 29v; on fols. 30v–84v this foliation continues in arabic numerals. A more recent recto-side sequence of arabic numerals appears on fols. 1–127 (discounting the numbering on the flyleaves and the new index on fols. 128–31). Folios 120–24 are now lost.

5. Despite the double repertory and the variety of scribes on fols. 1–88, the paper used for parts 1 and 2 of the source forms a single unit (that is, the idea of chronologically distinct layers of compilation is not supported by paper types); according to von Fischer, this is evident in the following distribution of watermarks:[12]

- Basilisk: fols. 1–34, 42, 48, 49, 51, 55, 59, 66–71, 80–83.
- Mount: fols. 35–40, 53, 57.
- Bell: fols. 46, 50, 85, 86.
- Arc: fols. 72, 73, 78.

The final section stands apart from the rest by virtue of the hardness of its paper and the crown watermark it carries.

6. In addition, fols. 1–88 feature eight music staves per folio; fols. 89–119, seven staves. Red ink was used to rule the staves of gatherings 2 and 3, and variations in the writing space coincide with the three sections: gatherings 1–5 = 19.6 × 17.5 cm; gatherings 6–7 = 21.3 × 18 cm; and the final section of the manuscript (gatherings 8 and 9) = 20.2 × 15 cm.

7. Von Fischer concluded from these data that subsequent to the initial compilation of gatherings 1–7, Scribe D completed gathering 5, made additions to gatherings 1–5 and 6–7, and brought his work to a close by binding the gatherings together, adding the paper of gatherings 8 and 9, and compiling an index at the end of the collection—a listing of the French repertory copied by himself and Scribe E.

Von Fischer's analysis of the scribes and the history of compilation were strongly contested by Nigel Wilkins in his 1964 dissertation on the Reina Codex.[13] Although the substance of Wilkins's arguments was published one year prior to the completion of the dissertation,[14] certain elaborations were developed in the unpublished dissertation and will be included in the condensed description below:

1. The scribes: von Fischer's Hands A and E are identical. This scribe, now labeled Scribe I, **alone** undertook the compilation and redaction of this collection of North-Italian and French secular song (details of Wilkins's comparison of scribes will be taken up shortly in my own description of the hands). Von Fischer's Hands C and D are also alleged to be identical, Wilkins citing—although with caution—ink colors as evidence. This scribe, now labeled Scribe III, was more at ease with French-texted compositions than was Scribe I. Wilkins does, however, present examples of corruptions in the French texts copied by Scribe III, and there he credits "the difficulty in reading the small and very abbreviated handwriting."[15] Von Fischer's Hand B remains for Wilkins a legitimately independent scribe, and is labeled Scribe II. Scribe IV (von Fischer's Hand F), the compiler of the fifteenth-century supplement at the back of the collection, was also responsible for the index. Wilkins argues that the index had clearly been created by someone unfamiliar with the compositions he was citing, thereby eliminating Scribe III as a possibility. The index hand, in fact, "corresponds exactly, as one would expect, considering the position of the folios, to that of Scribe IV."[16] Finally, Wilkins admits that various other hands, not easily distinguished, "seem to have dabbled here and there, at least in the writing of texts (e.g., fols. 35v–37)." This also includes the corrections and additions to the texts on folios 65r and 70r (not entered by Scribe III, as von Fischer has suggested).

2. This very different assessment of scribal hands prompted Wilkins to reconstruct the relative chronology of the manuscript's compilation, correcting von Fischer's earlier description.[17] According to Wilkins, Scribe I assembled a codex of seven gatherings consisting of blank paper produced in Italy at the end of the fourteenth century. He then went about filling selected gatherings and parts of gatherings, copying Italian works into the first three gatherings, most of the fourth, and the first four folios of the fifth. He reserved space in gathering 4 (fols. 40–42 and 44v–47) for additional Italian works, presumably to be entered at a later date. He also filled the sixth and seventh gatherings with French works; this could have taken place at a later date, for the format of the writing space changes in part 2 of the codex. In support of the thesis that one scribe

alone assembled the folios for both parts 1 and 2 of the collection, Wilkins notes that various points in the binding have been strengthened by strips cut from "an older parchment manuscript" (fols. 62v–63 [on 63], 67v–68, 72v–73 and 78v–79) as well as from still other sources (for fols. 6v–7, 18v–19, 48v–49, 55v–56 and 62v–63 [on 62v]). With the assembled gatherings before him, a new hand, Scribe II, added pieces to gathering 4, and other unspecified hands "dabbled here and there," until, finally, Scribe III (a Frenchman) "filled-in" gathering 5 with French works, gatherings 1–4 with French and Italian works (fols. 12v–13, 44v–45, and 46v–47), and gatherings 6 and 7 with other French compositions (fols. 65v–66, 72v–73, 77v). Scribe II, very likely, also added the music to folios 85r and 85v (presumably blank but already inserted into the manuscript as it was then constituted?). Part 3 of the codex, gatherings 8 and 9 (fols. 89v–118v) and the beginning of a tenth gathering (fols. 119–27), were then added to the earlier seven gatherings by a Scribe IV (a Frenchman). He drew his own red initials and most probably went back and contributed those on fols. 1v–2 and 77v, in gatherings 1 and 7 respectively. He completed his work by creating an index on fols. 126v–127 of works copied by Scribes I and III, and probably also listed Italian works and his own repertory on either the missing folios 120–24 or on others (128ff.) also now lost.

3. In concluding, Wilkins takes up the question of watermarks, but not so much to reexamine them as to support his already-formed ideas about the history of compilation. Adopting von Fischer's watermark description, Wilkins suggests that Scribe I had assembled and copied Parts 1 and 2 of the codex by 1398, using Basilisk paper for gatherings 1–3, and a "mixture" of Mount, Bell, Arc, and the same Basilisk paper for gatherings 4–7 of "his" codex, and adds that it would have been unlikely for von Fischer's Hands A and E to have "independently produced a similar mixture of paper, even though the Mount is absent from fascicles 6 and 7."[18] Working at a later date, Scribe IV used different paper for Part 3 of the collection.

A fresh investigation of the Reina Codex and the conclusions deriving from it will bear directly on the conflicting and/or inconclusive aspects of the analyses offered above; namely, those regarding the description of papers and gathering structure, examination of copyists's traits, sequence of compilation, and observations on the nature of the readings produced. The order in which the evidence is to be set forth in the present analysis corresponds to an investigative sequence in which decisions concerning scribal practices, variant readings, and stemmatic interrelationships rest as firmly as possible on the distinguishing of scribal hands. Analysis of scribal contributions depends, in turn, upon proper and thorough observation of every aspect of the copying process: pricking, ruling, format, layout, foliation(s), binding, and variances in ink colors. And, finally, all concerns regarding what, how, and when the contents of a source were copied must stand on an appreciation and understanding of the materials upon which it was copied—in this case, paper.

Although, as the late Allan Stevenson has pointed out, the study of paper for the precise dating of individual documents has been "cursed with ambiguities of provenance, dispersion and use," the inferences drawn from an informed and organized study of watermark evidence can be a fruitful bibliographical tool.[19] In identifying and distinguishing the seven paper types of the Reina Codex (see diagram 1) care has been taken to (1) view and describe marks from the mould side of the paper,[20] (2) adopt a method of measurement following Stevenson's model,[21] (3) distinguish marks made on twin moulds (marked in diagram 1 with an asterisk) from those made on other pairs of moulds;[22] and, finally, (4) provide beta-radiographs of the marks (not including twins, except in the case of mark #5) as well as references to published tracings. In citing tracings closely resembling the Reina marks, I have taken into account the fact that the Reina Codex sheets (all used in folio format) measure at least 27.1 × 42.6 cm.[23] Jean Irigoin's suggestion that the distance covered by twenty laid lines be included with the description of each mark has been followed.[24]

Basilisks

The most striking find in regard to the watermarks is the discovery of two basilisk papers in the manuscript; the one in gatherings 6 and 7 is distinct from that in gatherings 1 through 5. The basilisk mark cited by von Fischer as present in both parts 1 and 2 of the source is the one reproduced as plate C of the introduction to the Briquet volumes,[25] and found, in turn, on paper written upon in Bologna, in 1390. Briquet's basilisk does not, however, closely resemble either of the Reina marks. In the following description, measurements were made from the top of the uppermost ear to

folio / watermark / scribe

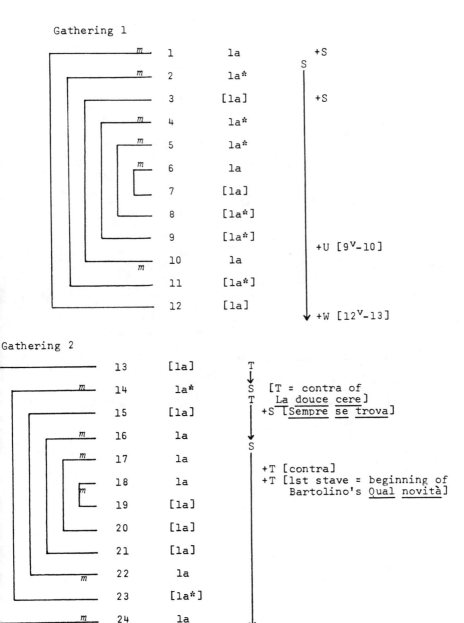

DIAGRAM 1. The Structure of Manuscript Paris, Bibliothèque Nationale, nouv. acq. fr. 6771 (Reina Codex). N.B. Numbers in brackets denote unmarked halves of bifolia. Asterisks indicate twin marks. *m* = mould side of paper.

Gathering 3

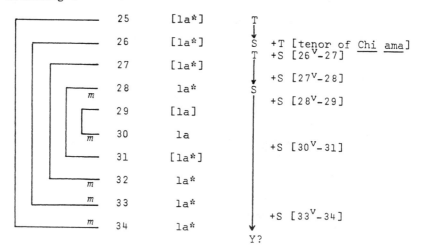

Gathering 4

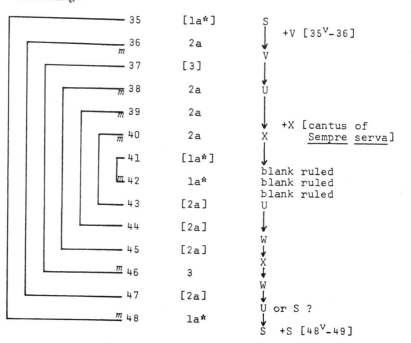

Gathering 5

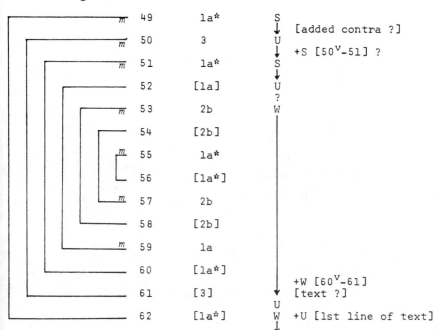

	49	1a*	S	[added contra ?]
	50	3	U	+S [50^V-51] ?
	51	1a*	S	
	52	[1a]	U	
	53	2b	? W	
	54	[2b]		
	55	1a*		
	56	[1a*]		
	57	2b		
	58	[2b]		
	59	1a		
	60	[1a*]		+W [60^V-61]
	61	[3]	U	[text ?]
	62	[1a*]	W	+U [1st line of text]

Gathering 6

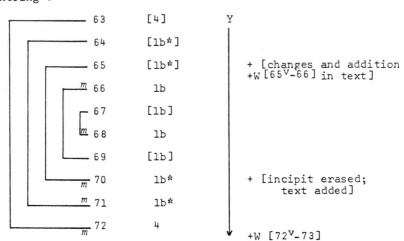

	63	[4]	Y	
	64	[1b*]		
	65	[1b*]		+ [changes and addition
	66	1b		+W [65^V-66] in text]
	67	[1b]		
	68	1b		
	69	[1b]		
	70	1b*		+ [incipit erased;
	71	1b*		text added]
	72	4		+W [72^V-73]

Gathering 7

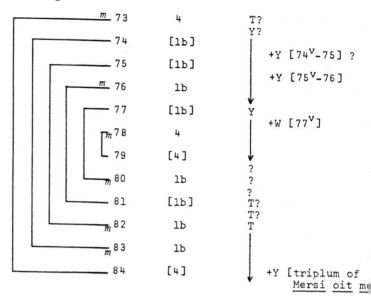

m 73	4	T?	
		Y?	
74	[1b]		+Y [74^v-75] ?
75	[1b]		+Y [75^v-76]
m 76	1b		
77	[1b]	Y	+W [77^v]
m 78	4		
79	[4]		
m 80	1b	?	
		?	
81	[1b]	T?	
m 82	1b	T?	
		T	
m 83	1b		
84	[4]		+Y [triplum of Mersi oit m⟨

Inserted Folios

m 85	3	V
m 86	3	blank
		blank
87	[3]	blank
		blank
88	[3]	blank ruled
		blank ruled

Gathering 8

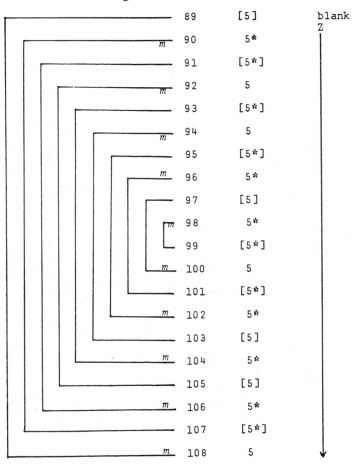

89	[5]	blank
90	5*	Z
91	[5*]	
92	5	
93	[5*]	
94	5	
95	[5*]	
96	5*	
97	[5]	
98	5*	
99	[5*]	
100	5	
101	[5*]	
102	5*	
103	[5]	
104	5*	
105	[5]	
106	5*	
107	[5*]	
108	5	

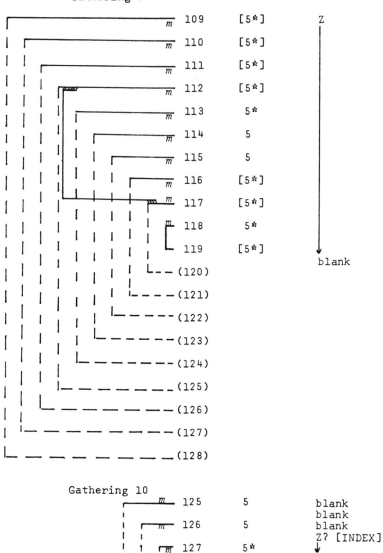

Gathering 9

_m_ 109	[5*]	Z
_m_ 110	[5*]	
_m_ 111	[5*]	
_m_ 112	[5*]	
_m_ 113	5*	
_m_ 114	5	
_m_ 115	5	
_m_ 116	[5*]	
_m_ 117	[5*]	
_m_ 118	5*	
119	[5*]	blank
(120)		
(121)		
(122)		
(123)		
(124)		
(125)		
(126)		
(127)		
(128)		

Gathering 10

_m_ 125	5	blank
		blank
_m_ 126	5	blank
		Z? [INDEX]
_m_ 127	5*	
		blank

the bottom of the rear foot, and across at the widest point—either from the front of the chest to the last wing, or from the chest to the rump. **Type 1a:** fols. 1, 2, 4, 5, 6, 10, 14, 16, 17, 18, 22, 24, 28, 30, 32, 33, 34, 42, 48, 49, 51, 55, and 59. **Type 1b:** fols. 66, 68, 70, 71, 76, 80, 82, and 83.

> **1a**—basilisk (plate 1; cf. Briquet 2660 [Ferrara, 1392], Mošin and Traljić 1064 [Venice, Reggio Emilia, Bologna, Lucca, Palermo, 1390–93]).[26] 57.5 × 15 [14/22.5] 4.5 mm. 20 laid lines = approx. 28 mm.
> **1b**—basilisk (plate 2; cf. Briquet 2632 [Ferrara, 1390]). 63 × 8 [19/18] 11 mm. 20 laid lines = approx. 24.5 − 25 mm.

Mounts in Circles

Here von Fischer cited Briquet 11890 as a match, and he reported a date of 1390 when, in fact, Briquet gives 1399. Again, I have distinguished two types, 2a and 2b. Not having been able to detect a twin for type 2a, I suspect that there is a possibility that type 2b is a twin mark of 2a. Be this as it may, they are used separately in the manuscript (2a in gathering 4 and 2b in gathering 5). **Type 2a:** fols. 36, 38, 39, and 40. **Type 2b:** fols. 53 and 57.

> **2a**—mount in circle, with a cross 28 mm above the top of the circle (plate 3; cf. Briquet 11854 [Lucca, 1388], Mošin and Traljić 6432 [30 × 43 cm. Dubrovnik, 1394 and Fabriano, 1398]). 26 × 15[13/15] 11.5 mm. diameter of circle = 40.5–41 mm. 20 laid lines = 26–27.5 mm.

> **2b**—mount in circle, with a cross 50 mm above the top of the circle (plate 4; cf. Briquet 11890 [Siena, 1399; here the sheet size appears to be very large—41.5 × 59 cm] = Mošin and Traljić 6437 [Siena, 1399; Fabriano, 1393–95] = Zonghi[27] 1291 [Fabriano, 1393; here the cross does not match]). 27.5 × 14[14/14] 14 mm. diameter of circle = 41 mm. 20 laid lines = 26 mm.

Bell

It is not clear whether von Fischer was citing Briquet 3953 when he offered the date of 1379 for the bell watermark. In fact, none of the published tracings matches the Reina Codex bell mark. Type 3: fols. 46, 50, 85, and 86.

PLATE 1. Reina Codex, watermark type 1a (fol. 1ʳ).

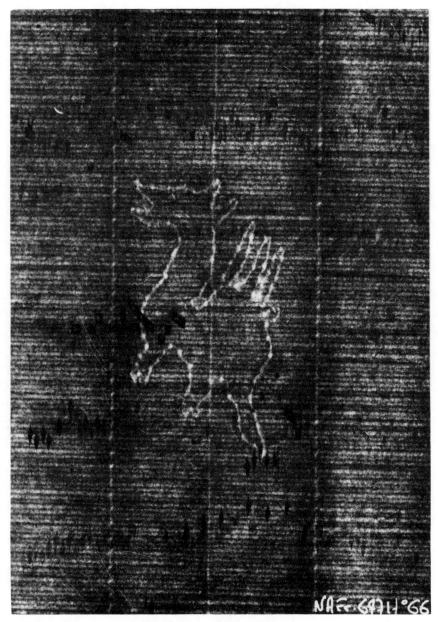

PLATE 2. Reina Codex, watermark type 1b (fol. 66r).

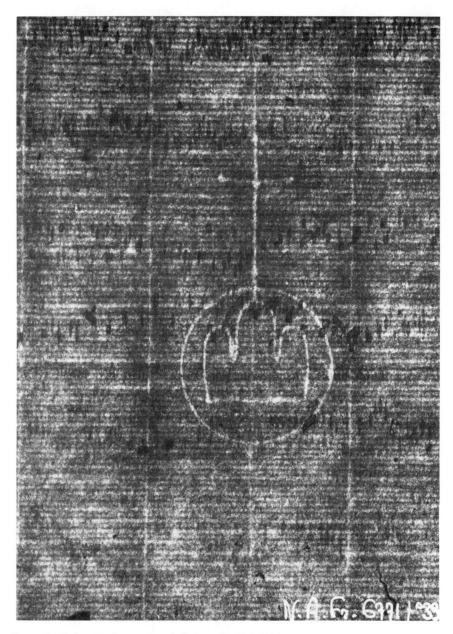

PLATE 3. Reina Codex, watermark type 2a (fol. 39v).

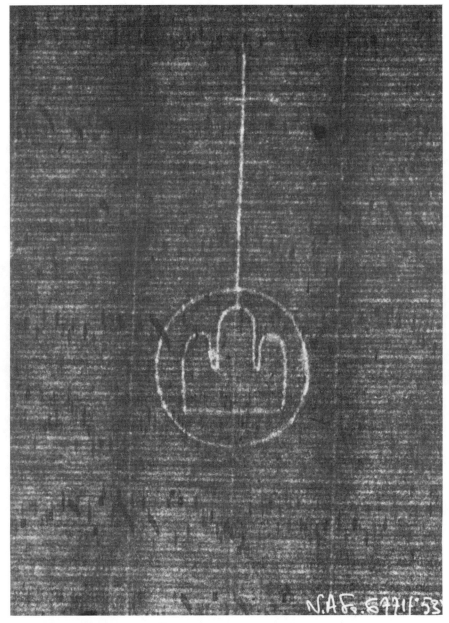

PLATE 4. Reina Codex, watermark type 2b (fol. 53v).

3—bell (plate 5; the state of the mark in the manuscript shows a broken chain line through the middle of the bell). 48 × 16 [24.5] 16 mm. 20 laid lines = 27.5–28 mm.

Arc

Von Fischer cited Briquet 788 (Fano, 1380). **Type 4**: fols. 72, 73, and 78.

4—arc (plate 6; cf. Briquet 791 [Lucca, 1393]; Briquet 799 [Paris 1406, va. ident. Turin 1410–11] should perhaps be eliminated, for the untrimmed sheet size is quite large—39.5 × 58 cm; Mošin and Traljić 431 [Syracusa, 1390]. 64 × 0 [28/26.5] 0 mm. 20 laid lines = 28.5 mm.

Crown

Von Fischer cited Briquet 4614 (Venice, 1387). The Reina mark does not closely resemble that tracing, but it is perhaps the best comparison possible.[28] I give the dimensions of the twin mark (Type 5*), and include a photograph for reasons that will become clear in the discussion of the makeup of gatherings 8 and 9. **Type 5**: fols. 92, 94, 100, 108, 114, 115, 125, and 126. **Type 5***: fols. 90, 96, 98, 102, 104, 106, 113, 118, and 127.

5—crown (plate 7; cf. Briquet 4614 [Venice, 1387; va. ident. Udine, 1396] = Mošin and Traljić 3243 [Venice, 1387; Udine, 1396; vs. Provence, Bourges, 1404–07]). 48 × 3 [27/26.5] 3 mm. 20 laid lines = 27.5–28 mm.
5* (twin)—crown (plate 8) 46 × 3 [26/26.5] 4 mm. 20 laid lines = 27.5–28 mm.

Notes on the Gatherings and Inserted Folios

Gathering 1

Rastrum: 18.5 mm, light brown ink.[29]
Margins: double lines, red ink.
Writing space: approximately 194–194.5 × 174–174.5 mm (varies, depending on distance between staves).
Pricking: matching where visible: fols. 1–6.[30]

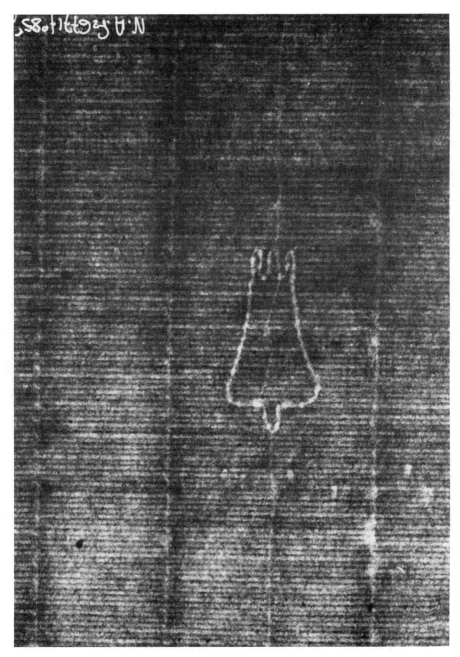

P<small>LATE</small> 5. Reina Codex, watermark type 3 (fol. 85ᵛ).

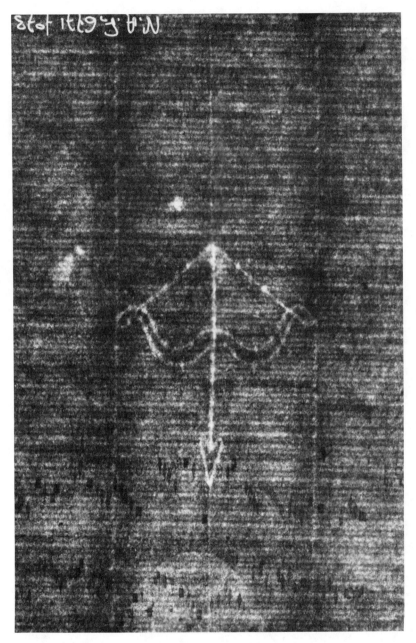

PLATE 6. Reina Codex, watermark type 4 (fol. 73r).

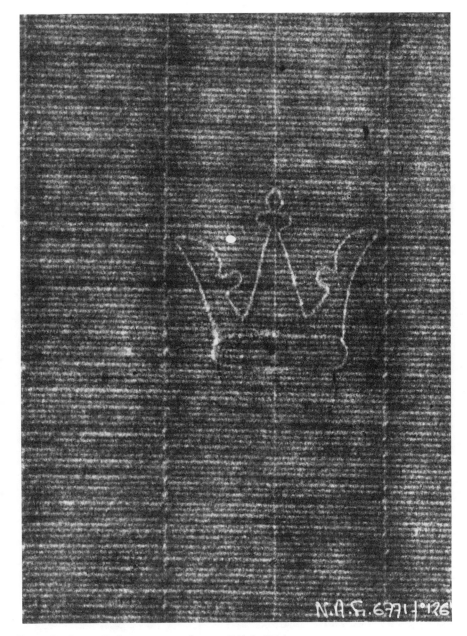

PLATE 7. Reina Codex, watermark type 5 (fol. 126ᵛ).

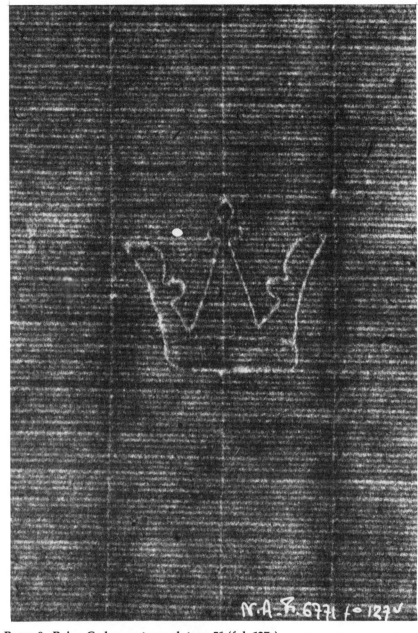

PLATE 8. Reina Codex, watermark type 5* (fol. 127ᵛ).

Foliation: recto—arabic numbers with semicircle, grey-brown ink (fol. 6 lacks the semicircle). Verso—roman numerals visible starting with fol. 8v (a bit of the descender *j* can be seen on fol. 3v).

Binding strips: parchment-on-paper strips at the center of bifolia 1/12 and 6/7, added after copying.[31]

Text and music: dark brown ink.

Gathering 2

Rastrum: 18–18.5 mm, red ink.[32]

Margins: double lines, red ink.

Writing space: approximately that of gathering 1.

Pricking: although some pricking on bifolium 18/19 seems to match surrounding bifolia (little of the pricking is visible), the area ruled for music is somewhat shorter in height than in the rest of the gathering; the height of the bifolium itself is approximately 2 mm shorter (this is visible even on microfilm). Bifolium 18/19 could be an addition or replacement.

Foliation: as in gathering 1; number on fol. 17r lacks the semicircle.

Binding strips: difficult to determine if strips are of parchment or paper. It appears that paper was used for bifolium 13/24, and parchment for inner bifolium 18/19; both were put in place after copying.

Text and music: layers can be distinguished by colors of ink: light brown ink on fols. 13r–13v, top of fol. 14r, *contra* on 17v, first line on 18r; dark brown ink on 14r, the addition on 15r, 18r, 19v–24v; light brown ink on fols. 16v–19.

Gathering 3

Rastrum: 18–18.5 mm, red ink. Use of a broader-nibbed pen than in gatherings 1 and 2. An additional line was drawn beneath the eighth stave, and it runs the width of the paper; it is a guide line for text, which in gatherings 1 and 2 had been placed on the lowest line of the eighth stave.[33]

Margins: double lines, red ink.

Writing space: as a result of the added text line under the eighth stave, height increases by approximately 7 mm: 202–203.5 × 173 mm.

Pricking: matching for the entire gathering (where visible). An extra prickmark is occasionally visible in the outer margin; it was used to align the text guide line under the eighth stave.

Foliation: roman numerals stop on fol. 29v; arabic numbers are used on fols. 30vff (a different hand from that on the recto sides). Arabic numbers on recto sides lack semicircles on fols. 25, 26, 28, and 29. The tail of a *custos* covers the number on fol. 30r, suggesting that the foliation of the entire gathering was completed before copying resumed on fol. 30r or 31r.

Binding strips: bifolium 25/34 is reinforced with a paper strip; bifolium 29/20, with parchment. Both strips were put in place after copying.

Text and music: dark brown ink on fols. 26r, 28r–34r; light brown ink on fols. 25r, 25v, 26v, 27r, 27v, and 34v. The composer attribution on fol. 25r ("Dompni pauli") appears in the ink of the rest of the folio; those on fols. 25v and 26v are in red ink. Long note stems on the first stave of fols. 28r, 28v, 29r, 29v, 30r, 32v, and 33r.

Gathering 4

Rastrum: 19 mm, light brown ink (nearly transparent). Use of a thin-nibbed instrument.

Margins: double lines, brown ink: single-line right margins on bifolium 37/46.

Writing space: approximately 193.5–195.5 × 173.5–174 mm for bifolium 35/48; 197 × 174 mm for bifolia 36/47, 38/45, 39/44, and 40/43; 197–197.5 × 179–182 mm for bifolium 37/46; 195.5–196.5 × 174.5–175 mm for bifolium 41/42.

Pricking: often all eight prickmarks in the outer margins are visible. The following bifolia match: 36/47, 38/45, 39/44, 40/43, and 41/42. Bifolia 35/48 and 37/46 show evidence of independent origins, for they match neither each other nor the rest of the gathering.

Foliation: continuation from gathering 3. A new hand in foliation numbers is clear by fol. 43r (note the different form of 3). Beginning with fol. 39v, one finds a clear offsetting of recto-side numbers onto the verso sides (caused by use of a different ink?). Blank-ruled fol. 42r carries a folio number. A *custos* appears to cover the folio number on 35r.

Binding strips: bifolium 35/48 strengthened with a paper strip put in place before copying. Parchment strips (a short one added to a longer one to cover the entire height of paper)

were used for bifolium 41/42. Pricking on strips indicates that
these are the remains of trimmed manuscripts.

Text and music: a very dark, brown-black ink (almost black)
from fols. 38r to the top of 39v and from 47v to 48v (the ink on
47v is darker than that on 48r); relatively dark brown ink is to
be found on fols. 35r–36r, the bottom of 35v–37v, 43–44, and
44v–45; a light, transparent brown ink is found from fols. 39v
(bottom) to 41r, and again on fols. 45v–46; a light brown ink
(somewhat darker than that of fols. 39v–41) appears on fols.
46v–47.

Gathering 5

Rastrum: 19 mm, brown ink. Bifolium 50/61 is ruled in a light
brown ink.

Margins: double lines, brown ink. Bifolium 50/61 has single-line
margins in light brown ink.

Writing space: approximately 195 × 174–175 mm; bifolium 50/61
= 195–197 × 178.5–181 mm.

Pricking: very little is visible.

Foliation: continuation of foliation in gathering 4 (the different
forms of the number 5 on the recto and verso sides of folio 50
attest to the independence of the two sequences). Starting
with fol. 51r, one finds a new hand and a rust-colored, light
brown ink; no further use of semicircles. Verso-side foliation
appears in a darker ink, starting on fol. 50v. Numbers on fols.
53–55 (recto sides) have been retraced by a modern hand. Ink
color, hand, and the location on the page of verso-side foliation changes at 57v.

Binding strips: parchment strip strengthening bifolium 49/62
was put in place before copying. Parchment strips (a long and
short one to cover the entire height of the paper) put in place
before copying on bifolium 55/56.

Text and music: 49–51v, brown ink; a very dark, brown-black ink
on fols. 52r–52v (close to that of fol. 47v); a change to a
lighter, grey-brown ink on fol. 53r (resembling that of fols.
44v–45).

Gathering 6

Rastrum: 19 mm, very light brown ink (resembling that of
gathering 4). Contrary to common practice in the rest of the

manuscript, fols. 66v and 67r (non-conjugates) were ruled together.

Margins: double lines, same color ink as the staves.

Writing space: 211–213.5 × 177–182 mm (space between staves here averages 9–9.5 mm versus 6–7 mm of gatherings 1–5).

Pricking: bifolium 64/72 contains no visible pricking; other bifolia match each other.

Foliation: beginning on fol. 62v (with the exception of fol. 63v) the verso-side numbers are accompanied by a dot or dash on either side. The recto-side numbers continue in the light, rust-colored brown sequence of gathering 5.

Binding strips: parchment strips (one long, one short) are present on bifolium 63/72, of which the lower short strip is now missing on fol. 63r; these strips were evidently taken from a manuscript with four columns of text and red capitals. The same source provided the strips used for bifolium 67/68. Both sets of strips were put in place after copying.

Text and music: light brown ink was employed for the entire gathering. Changes and additions to the text on fols. 65r and 70r are also in a light brown ink. Additions on fols. 65v–66 and 72v–73 are in black ink. Red notes were originally brown.

Gathering 7

Rastrum: 19 mm, light brown ink.

Margins: double lines, light brown ink.

Writing space: 211–213 × 178–182 mm. Bifolium 73/84 departs from this in that fols. 73v and 84r appear to have a stave ruled at the very bottom of the leaf.

Pricking: bifolia 74/83, 75/82, 76/81, and 77/80 contain matching prickholes (nine pricks in the outer margin, starting at the top of the first stave; this contrasts with the eight holes in gatherings 1–5, each corresponding with the bottom line of a stave); bifolia 73/84 and 78/79 have no visible pricking.

Foliation: continuation of gathering 6. Starting on fol. 75r, the recto-side numbers have been retraced in pencil by a modern hand (excepting numbers 77, 82, and the 4 of 84). A different hand appears to have numbered folio 77r. Continuation of verso-side foliation.

Binding strips: those on bifolia 73/84 (now missing on 84v) and 78/79 were taken from the same source that served for gathering 6. All were put in place after copying.

Text and music: very light brown ink on fol. 73r; brown ink for fols. 73vff; addition on fol. 77v entered in black ink; a slightly darker ink is to be found on fols. 81vff. Red notes were originally brown.

Inserted Folios 85–88

Rastrum: 18.5 mm on bifolium 85/88; 86/87 is blank, with only the vertical margins drawn in; brown ink.
Margins: single lines, brown ink.
Writing space: 210–211.5 × 180–184 mm.
Pricking: the two bifolia match.
Foliation: 85r–119 (recto side) in the same hand as fols. 51r–84. No verso-side foliation from fol. 85 to the end of the manuscript.

Gathering 8

Rastrum: approximately 18.75–19 mm, light brown ink. Seven staves per folio. Text guide lines are drawn in pencil (sometimes lacking under the lowest stave). Because of space reserved for the initial capital letter, fol. 89v clearly was meant to be the initial page of this part of the collection.
Margins: single lines, pencil.
Writing space: approximately 202 × 150 mm.
Pricking: entire gathering is uniform in this respect; double pricking is evident on folios 102–108.
Foliation: continuation of recto-side numbering from gathering 7 and inserted folios 85–88. No verso-side numbering.
Binding strips: modern white paper strips have been added between fols. 89v–90, 90v–91, 92v–93, 105v–106, and 107v–108. It is possible to establish conjugacies through watermark analysis and the measurement of chain lines on unmarked folios (see observations for gathering 9). An old paper strip, together with a modern one, strengthens the outer bifolium of the gathering, 89/108; it was put in place after copying. An old parchment strip appears at the center of the gathering, reinforcing bifolium 98/99; it may well have been added after the copying of the music, but was already in place by the time the capital letter *S* was drawn on fol. 99r.

Gatherings 9 & 10

Rastrum: approximately 18.75–19 mm, light brown ink. Seven staves per folio. No patterns could be detected in the slight variances of stave heights that would help clarify the gathering structure. Text guide lines are drawn in pencil (sometimes lacking under the lowest stave).

Margins: single lines, in pencil.

Writing space: approximately 202 × 150 mm.

Pricking: uniform for fols. 109–119; fols. 125–127 match (that is, they were originally pricked for music, as was gathering 9; they were subsequently repricked for use as an index for the drawing of double and triple vertical guide lines). Folio 119v was not ruled for music.

Foliation: a continuation of the recto-side numbering of gathering 8. The number 120 can be seen offset onto fol. 119v. Likewise, a number 128 was offset onto the upper left corner of fol. 127v (N.B. it does not match the form of the modern number 128 on the bifolium flyleaf 128/129). Foliation on the rectos of 125, 126, and 127 differs in ink color (a nearly transparent grey-brown) from that on fols. 109–119.

Binding strips: modern white paper strips (attached with thin, white thread) join the following folios: 109v–110r, 110v–111r, 113r and 116v, 114r and 115v. On the mould sides of fols. 112 and 117 there is an older, dark-brown paper strip that was glued into place after copying (it is included in diagram 1). Fols. 117 and 118 had been glued together directly, without the use of a strip. Older (original?) thick thread is visible between folios 118v and 119r, and the glue that must at one time have held some type of reinforcing strip between fols. 118v and 119r still remains.

Text and music: no red capitals on fols. 115v–119r; solid red capitals on fols. 89v–96r, 99v–107r; void red capitals on fols. 96v–99r, 107v–115r.

Taken together, this evidence suggests a clear view of the structure of the ninth and what remains of the tenth gatherings (see diagram 1). Most important in determining conjugacy—and in dispelling the confusion created by the modern (and older) binding strips detailed above—are the identification of the folios carrying the twin mark (Type 5*), measurement of chain lines in the clearly visible conjugacies of gathering 8, and the detection of the mould sides of all unmarked folios. In short, folios 125–127 cannot be made to "fit" with presumed conjugates among folios 109–119.

Corroborating evidence that aids in arriving at the present structure includes the musical continuity on the versos and rectos of all openings from fol. 109–119 (excepting fols. 112v–113), the fact that fol. 119v had not been ruled for music, and the change in hand and ink color for folio numbers 125–127. Evidently, gathering 8 at the outset had been accompanied by a structurally similar gathering, both of which were composed of ten bifolia; and, though foliation probably ran past 119, only folios 109–119 were perhaps ruled and used for the copying of chansons (the absence of capitals on fols. 115v–119r indicating that these folios were the last to be filled, at a slightly later date than the others). Sometime after the creation of the index on then unnumbered folios, someone removed the presumably blank fols. 125–128. Evidence of the longstanding precariousness of the resulting unjoined leaves lies in the existence of an old paper strip between fols. 112 and 117, as well as the glue between fols. 118 and 119. When the index folios were foliated at an even later date, the numbering took up from the last numbered folio, no. 124. Since then—but before the manuscript entered the Bibliothèque Nationale in Paris—other blank folios have been lost: the original fols. 120–124, and at least three others after the present folio 127. Two questions remain unresolved: why was folio 119v not ruled along with 118r, and why did blank folio 125 survive (or why was the index placed on the folios it occupies)?

Scribes

Distinguishing among the scribes can be the most difficult of the tasks required in the examination of a manuscript in which a number of copyists have participated. Because the identification and sorting of music and text scripts rests partly on subjective criteria, it is of vital importance that the more objective exercises in observation be carried out first. I do not pretend to have satisfactorily solved all of the problems posed by the scribal contributions in the Reina Codex—quite the contrary, new questions are raised and not a few old ones are left unanswered—but I have made an effort to draw a picture that at least in many respects accords with the "harder," more objective evidence already presented (i.e., the watermarks and additional codicological notes given above).[34]

In his reply to Nigel Wilkins's revised description of the Reina Codex, Professor von Fischer made the point that his colleague's proposed identity of Hands A and E—as seen in a comparison of parts 1 and 2 of the manuscript with regard, among other things, to the drawing of final notes and the decoration of capital letters—

was not as clear as Wilkins would have us believe.[35] I would echo this reservation by von Fischer and note that a manuscript-wide examination of final notes and *finis punctorum* divisions provides us with no clear-cut coincidences of scribes and manner of decoration; in this respect, scribes, quite possibly, can imitate each other, or their exemplars. This concern carries over into Wilkins's question of "wavy" decorations in capital letters, for one finds, as an example, such an embellishment in a capital *T* on fol. 40r (a leaf, by Wilkins's own admission, demonstrably not the work of his Scribe I). Moreover, the comparison by Wilkins of von Fischer's Hands A and E in "random" folios 27v–28 and 63v–64 likewise remains unconvincing. Despite a general resemblance of script, I note the following discrepancies:

1. F-clefs differ: those on fols. 63v–64 are attached to the extreme left vertical margin, while those on 27v–28 are not.
2. The size of the script is significantly larger in gathering 6 (especially evident in the drawing of *custodes*).
3. Final double bars occur in pairs in the earlier gatherings; an odd number of such bar lines are common in the work in gatherings 6 and 7.
4. Capital *O*'s are not identical (to use Wilkins's term), nor are miniscule *g*'s (two forms of which appear on the very same folio, 27v), *d*'s, capital *S*'s, and *E*'s.

A further link, Wilkins contends, between the two parts of the collection copied by von Fischer's hands A and E is created by what must be a single scribe's access to red ink—used in the ruling of staves in gatherings 2 and 3 of the Italian section, and for coloration in gatherings 6 and 7. This, too, cannot pass as incontrovertible proof, for (1) the ruling of gatherings is part of the preparatory procedure in the compilation of a manuscript and could easily have been accomplished by hands other than the ones responsible for the copying of music and text, and (2) the pieces employing red ink are confined to the centers of both gatherings 6 and 7, and almost certainly in gathering 7 may thus constitute a distinct layer inserted into the collection.[36]

The arguments in the present evaluation have nothing directly to do with a scribe's nationality or native language; in fact, I agree with Wilkins that textual corruptions point to the participation of an Italian scribe in von Fischer's part 2 of the collection.[37] What is at issue here is the concept of identity that Wilkins espouses so strongly. Could it not be that a different, yet still Italian, scribe was

responsible for part 2 of the collection? I suspect that the tempta-
tion to posit a theory of identity arose partly because the final
folios of gathering 7 do, in fact, contain the work of a scribe found
in the earlier gatherings (see below, Scribe T). The following notes
on Scribes S–Y, as distinguished in the present writer's analysis,
do not pretend to be exhaustive, and treat only of those features
that have aided in isolating the work of a particular scribe. Somewhat more attention has been allotted here to music scripts than
had been done by Wilkins.[38]

Scribe S

> Music: a large, clear hand; generally long straight *custodes;* large
> parallelogram-shaped sharp signs with four dots in the cor-
> ners;[39] tails of semiminim and triplet flags curve downward;
> in F-clefs, only the single rhomboid has a tail; use of paired
> divisional and final bars; final, broad-peaked, accordion-
> shaped notes. He experiments with heightened stems, space
> permitting (gathering 3).
>
> Text: a large, clear hand;[40] experiments with forms of capital
> letters, and one case of extended descenders (fol. 9); use of a
> bold, mordentlike abbreviation for the letter *r;* varied use of
> final punctuation dots, often three in number, in a pyramid
> shape, and occasionally using four or more in ellipsislike
> fashion; long *s.*

Scribe T

> Music: in F-clefs, both the single and double rhomboid shapes
> are drawn with stems; tails of semiminim flags do not always
> curve downward and are often drawn as straight lines point-
> ing up at a 45-degree angle away from the note stem;[41] less
> boldly drawn sharp signs and some experimentation with the
> shape of the parallelogram sides (fols. 13v, 82r); divisional
> and final bars are drawn in pairs (sometimes "tied" together
> in panpipe fashion) as well as three single strokes.
>
> Text: two forms of miniscule *g,* one with a closed, looped tail,
> and the other with a longer, "open" tail extending to the left.

Scribe U

> Music: a similar but smaller, more spidery hand than Scribe S's;
> many of the sharp signs lack dots, or at least the full comple-
> ment of four.[42]

Text: long, sloping ascender on miniscule *d* (sometimes looping around to the right); exaggerated descender on the long final *s* (descenders sometimes added to miniscule *f* and *i*); the scribe may be associated with the large, characteristic "Tenor" and "Contratenor" designations sometimes used for the lower voices in three-part compositions (see the added "Tenor" on fol. 20v); long, extended cedillas.

Scribe V

Music: use of the imperfect breve form for *custodes*;[43] triplet flags with long tails extending to the right (some parallel to the stave); sharp signs contain four dots, but they are placed in the middle of the four sides of the parallelogram; inconsistently shaped C-clefs; when present, very tight, jagged, accordionlike final notes.

Text: some use of miniscule *g* with looped tail extending to the left; elaborated cedillas (three-stroke form).

Scribe W

Music: thin, spidery hand with pointed semibreves; sharp signs often lack dots; few accordion-shaped final notes; F-clefs often lack a stem on the single rhomboid, and sometimes lack stems altogether; note stems incline to the right; semiminim flags are drawn to the right with short tails curved downward; tendency to prepare more C-clefs than are needed, some of which are then transformed into F-clefs by the addition of stems (fols. 45r and 59r).[44] Use of an extremely sharp pen nib which slightly tears the paper in the process of writing, often creating lateral "bleeding" around the vertical strokes of final bars.

Text: cursive features; thick, long *s* and *f*; looped ascenders for *b*, *l*, and *d*.

Scribe X

Music: resembles Scribe V in the drawing of final notes, sharp signs and triplet flags; thick-bodied *custodes*; B-flat in signature has the shape of a C-clef with added upper and lower stems (fols 40v and 45v).

Text: a striking hand, with looped *l* and *b*, long sloping ascender in miniscule *d* (often elongated and curving back to the right in a thin stroke); exaggerated tail on the long *s* and a long curved limb in miniscule *h;* use of single and double diagonal strokes to mark sections and endings of texts.[45]

Scribe Y

Music: a bold, clear hand; C- and F-clefs are consistently attached to the left outer margin of fols. 63r–72v and 73v–80v; F-clefs are generally drawn with two tails; fols. 74r, 74v, 75r, and 75v exhibit "transformed" C-clefs of the type mentioned in the description of Scribe W above; here, however, the effect is deliberate, for the clef is also found in the middle of line 2, fol. 75r; sharp signs sometimes lack dots (with some experimentation on fol. 66v); final bars seem to vary greatly, with an odd number of strokes favored; *custodes* vary considerably, including the use of curved tails. The script of fols. 73v–76v is similar to that of Scribe T, but smaller (however, note the use of single-stroke semiminim flags on fols. 73v and 81v [also present on fol. 43r]).

Text: use of miniscule *g* with a long tail extending to the left; large, elaborate cedillas; no long *s;* tops and bottoms of capital *S* are often "flattened." The hand responsible for the corrections and additions to the text on fols. 65r and 70r is unlike any other in the manuscript. It does, surprisingly enough, bear a close resemblance to the hand of a scribe who copied Landini's *Poy che da te convien* as a later addition in the fragmentary manuscript Grottaferrata E.β.XVI, fol. 3r.[46]

Compilation of the Collection

The repertorial significance of the Reina Codex stems, first of all, from its important collection of late 14th- and early 15th-century French—and even Flemish—*unica: virelais, ballades,* and *rondeaux* (part 2 of the MS). The Italian section (part 1) comprises works by well-known native composers (notably, Jacopo da Bologna, Giovanni da Cascia, Bartolino da Padova, and Francesco Landini), but it contrasts with contemporary Florentine collections in going beyond typical anthologizing tastes with its remarkable transmission of many anonymous North-Italian madrigals and ballatas, some displaying close ties to the unwritten tradition. Taken as a whole, then, the repertory of this collection suggests a northern, or at least non-Tuscan, musical center in which works by Florentine composers (principally Landini) were also available. There is reason to believe, in fact, that a reconstruction of the history of its preparation points specifically to Paduan circles as the most likely source for the codex itself.

Although gatherings 1, 2, and 3 form a unit, some question remains as to which of the first two was originally intended to open a collection that, in its infancy, appeared to be limited to North-Italian repertory—compositions by the masters Bartolino da Padova and Jacopo da Bologna. On folio 13v, the layout of *La douce cere* reserves considerable space for a large capital *L* (perhaps even an illuminated letter),[47] evidence perhaps that honor of first place had originally been intended for gathering 2—and northern Italy's foremost composer at the turn of the fifteenth century, Bartolino da Padova, whose works dominate the gathering. In the end, the Jacopo gathering was placed first, with the addition of red capitals to the three voice parts of *Sotto l'imperio* on opening 1v–2r.[48]

Scribes S and T appear to have been close collaborators, a relationship attested to by several factors: (1) they worked on identical paper; (2) the considerable internal complexities regarding the alternation of the two hands in gathering 2;[49] and (3) T's addition of a textless contratenor part to S's copying of Bartolino's *El no me giova* (fol. 17v).[50]

Gathering 3 opens with a ballata attributed by Scribe T to a composer who, judging from the manuscript dissemination of his works, was little known in northern Italy: Paolo da Firenze (Paolo Tenorista). Doubts cast on Paolo's authorship of *Perch' i' non seppi* (fol. 25r) are here reinforced, for this could be Scribe T's single contribution from the known Florentine repertory (the responsibility of which falls primarily on Scribes S, U, and V).[51] Gathering 3 continues with what may be the work of local Paduan composers—compositions by Jacobeli Bianchy, Joh. Baçus Correçarius, and an Henrici[52]—as well as a small collection of anonymous ballatas and madrigals. At least some of the ballatas have been recognized as belonging to a North-Italian tradition of *sicilane* "recast into ballata form."[53]

Another layer of copying is represented by Scribe S's addition of Landini, Giovanni, and Jacopo works to these three gatherings. Some of these works, those on folios prior to fol. 29v, were copied into available space at the foot of openings (as well as on blank-ruled fols. 1r and 12v); others, on fols. 30–34, occupy the major portion of the folio. All of these additions may have been entered at the same time.

The nature of the anthology underwent a change in gatherings 4 and 5, where one finds the use of new papers together with paper type 1a and the work of different scribes. A chronological sequence of compilation in which a unit made up of gatherings 1, 2, and 3 was followed by a second unit comprising gatherings 4 and 5 is supported by the following evidence:

1. Gatherings of one paper type are followed by gatherings of "mixed" paper types that include the first type.

2. The collaboration of only two scribes in the early gatherings precedes the more complex situation of later sections involving four new scribes working in conjunction with one from the first group.

3. The foliation (now looking ahead to include gatherings 6 and 7) is executed in "layers" that extend across gathering boundaries; in other words, changes in type, hands, and ink colors of the foliation numbers do not coincide with the beginnings and ends of gatherings.

4. The format of *La douce cere* and the red capitals of *Sotto l'imperio* suggest their intended location at the beginning of the collection.

5. Although binding strips were put in place before the copying of music in gatherings 4 and 5, it is doubtful, in light of the principles of repertorial arrangement customary in other Trecento sources, that the pieces on the opening rectos of these gatherings (e.g., the anonymous madrigal *Spesse fiate*) would have been intended to head the collection.

Gatherings 4 and 5 represent, then, a second layer of compilation in the collection: the folios share, in addition to a similar manner of preparation, similar paper types and scribes. They may originally have been compiled by Scribes S and U, with the other scribes contributing either on folios left blank-ruled by them, or on inserted or replacement folios. The contributions of Scribe S, the principal copyist of gatherings 1–3, continue to be offered on paper type 1a, but the different manner of paper preparation points to a new layer of work by him. Scribe V seems in particular to have been responsible for the insertion of paper type 3 in these gatherings (at least bifolium 37/46 and perhaps also 50/61). The folios now numbered 85–88, consisting of paper type 3 and, like fols. 35v–37v, carrying the work of Scribe V, may at one time have simply been appended to the end of the first five gatherings; the displacement of these folios to their present location perhaps took place after the compilation of gatherings 6 and 7. Finally, an association may be revealed between Scribes U and W, for on fol. 62r Scribe U copied the first line of text of the anonymous *ballade A gre d'amours, dame* (the remainder of text as well as the music are the work of Scribe W).

With regard to repertory, gatherings 4 and 5 continue the collection of anonymous ballatas, *siciliane,* and madrigals begun in gathering 3. In fact, the filling-in of gathering 3 (fols. 30v–34) may have taken place after gatherings 4 and 5 were begun, space having been reserved for Giovanni and Jacopo works in the earlier gathering. As it is, an even more heterogeneous mix resulted, including the addition of more anonymous works, two Landini ballatas, and further Bartolino compositions.[54] The only gathering that appears originally to have been intended for Landini's works is gathering 5, but his ballatas were added on earlier folios primarily by Scribes S and V.[55] However, less than half of the fifth gathering represents Landini; new Scribe W, who also contributed on fols. 44v–45 and 46v–47 (with both French and Italian works) began on fol. 53r what has been termed part 2 of the Reina Codex.[56] This section of the manuscript, which extends to folio 84v and contains primarily a fourteenth-century French repertory, does not form the single entity with gatherings 1–5 as reported by Wilkins. Instead, it stands apart from the first section for the following reasons: (1) it displays a different manner of preparation; (2) it comprises new paper types, 1b and 4, and altogether lacks the paper types of gatherings 1 through 5; and (3) gatherings 6 and 7 contain the work of a new scribe, Scribe Y, who may be largely responsible for the compilation of these folios. Only the presence of Scribe T at the very end of gathering 7 ties the fourteenth-century Italian and French sections of the source together.

Gatherings 6 and 7, although clearly forming a single, distinct unit in the history of the compilation, also display elements of internal complications, and thus may represent more than one layer of compilation. Most noticeable in this respect is the presence of red capitals on fols. 77v and 80r, the catchword at the bottom of folio 67v, and other added material in the extreme inner or outer bifolia of the gatherings.[57]

Finally, gatherings 8 and 9 (and the leaves representing the beginning of a tenth) were added much later to the collection. As shown earlier, these appear to have been uniformly prepared, despite the now-lost folios of gathering 9 and the confusion created by the subsequent addition of binding strips.

Although the purpose of this essay has been to describe and redress the physical aspects of the Reina Codex, it is appropriate to conclude this examination of the manuscript by presenting some observations on repertorial groupings and scribal contributions to the readings of the compositions.

Let us first consider the nature of Scribe V's work: three ballatas by Francesco Landini—*In somm' alteça* (fols. 35v–36), *Donna, che d'amor senta* (fol. 36v), *Non dò la colp' a te* (fol. 37r)—an unattributed ballata, *Ochi piançete* (fol. 37v), and two instrumental arrangements on fols. 85r and 85v.[58] Although Leo Schrade elected not to follow the Reina Codex readings for two of the three Landini ballatas in his critical edition of the composer's works, he nevertheless contended that the presence of Italian notational elements in these works (against the French practices in states found in concordant sources) signaled *by itself* their relative proximity to Landini's original conception.[59] However, the isolation of this scribe's work in the Reina Codex—including the instrumental arrangements on fols. 85r and 85v—might instead suggest that the notation of the ballatas may have more to do with Scribe V's training and/or personal preferences than with any intrinsic authority the readings with Italian notational traits might carry. *Donna che d'amor senta*, for example, is not transmitted in Schrade's "main" source (MS Panciatichi 26 of the Florence National Library), and he was thus tempted to give "preference to the version in Reina on account of the Italian notation, which is the most likely form of the original."[60] By so doing, however, Schrade introduced into his edition what may be a particular trait of Scribe V, for the repetition of notes in bar 12 of the tenor part (notated in fewer but longer notes in two other sources) appears to be symptomatic of his work.[61] The avoidance of certain longer note values is also to be found in one of the other Landini ballatas copied by Scribe V, a reading that Schrade decided not to follow, citing a number of errors in this transmission. He would, presumably, have incorporated the repeated notes in his critical edition had it not been for the presence of these other, erroneous readings.

In Landini's *Non dò la colp' a te*, Scribe V avoided the syncopation over-the-bar in measures 32–33 of the contratenor part (yet he incorporated a second one in measures 34–35). This is a particularly instructive example, for, as has been noted elsewhere, it was precisely at such a point of rhythmic complexity that Scribe V abandoned his instrumental arrangement of a French *virelai* on fol. 85v.[62] Two sides of one argument may be directed against Schrade's underlying "one main source" approach. First, notational language is clearly an area in which scribes could exercise initiative in the copying process, and, therefore, mensural practice and the interpretation of rhythmic complexities cannot be placed at the top of a hierarchy of criteria that establish authoritative readings.[63] Second, notational language (in the Reina Codex, Italian vs. French traits) is but one of the variables in the copying process, and thus it cannot ipso facto determine the authority of the other parameters involved: texting, ligaturing, and vocal scoring.

Another example of coincidence of scribe and specific copying traits in the transmission of compositions may be found in Scribe T's apparent preference for the distinctive Italian texture of texted cantus and tenor parts with a textless contratenor in Bartolino's three-voice compositions in gathering 2. In three of these—the madrigals *La douce cere* and *Alba colunba*, and the ballata *Senpre, dona, t'amay*—this scoring departs from that of concordant readings.[64] The ballata *El no me giova* (fol. 17v) was originally entered as a two-voice composition by Scribe S; to it, T added a textless contratenor part now thought to have been composed by Matteo da Perugia, and, as mentioned earlier, found alone in the Modena manuscript.

Further examples of scribal practices include the following: (1) the handling of syncopations (both within the bar and across-the-bar) in Jacopo's and Bartolino's works;[65] (2) Scribe S's incorporation of such syncopations in his Jacopo gathering (*Nel bel giardino* on fol. 5r, *O cieco mondo* on fol. 5v, and *Prima virtute* on fol. 6r),[66] as well as in other repertory (e.g., the anonymously attributed ballatas *Lasso per ben servire* on fol. 28v and *Donna nascosa* on fol. 32r, as well as *Se questa dea* on fol. 33r); (3) the deliberate changes introduced in the texts, particularly evident in the infusion of Venetian-Emilian dialectal traits into Tuscan texts;[67] (4) the less corrupt state of the compositions at the end of the seventh gathering, as compared to much of the rest of the French-texted repertory in gatherings 6 and 7.[68]

The work of the scribe (perhaps the "outsider" indicated on page 101) who corrected and added text to Machaut's *Gais et jolis* (fol. 65r) and the anonymous *ballade A discort son desir* (fol. 70r) is certainly editorial in nature. While the underlay may still be faulty on fol. 65r, much of it is not attributable to the new scribe, who clearly was making an attempt to adjust what was before him. And on fol. 70r, he was forced to work with a *ballade* in which the notes had already been "fixed" on the page without regard for proper spacing had text accompanied them. That the new scribe was concerned with proper alignment is shown in his use of coordinating lines between syllables and notes.

Thus the contributors to this collection cannot all necessarily be considered *mere* copyists, or, to use a more euphemistic equivalent, professional copyists. They were, in many respects, editors of the repertory they worked with. We should, as Margaret Bent has stated, be "grateful for the clarification by a musically skilled scribe of features such as text underlay,"[69] for even if gatherings 6 and 7 of the Reina Codex are not the ideal source for the French repertory of the late fourteenth century, we might still appreciate the fact that this section of the manuscript, or the earlier Italian section, may have transmitted certain features of a composition properly and others incorrectly. The editor's quest need not be for manuscripts that are on the whole more correct than others, but rather for individual readings—and even certain elements of readings, such as texting and vocal scoring—that are more correct. The full benefits of contemporary medieval scribal expertise can best be put to use (and identified in critical reports) only after we have analyzed scribal practices in the repertory under consideration.

A careful codicological and paleographical study of the great late-medieval anthologies of secular polyphony can form the proper basis for an analysis of the readings they contain. As shown in this study, far from detracting from its value as a collection, the complexities of structure and scribal contributions help us to view the Reina Codex, and other coeval anthologies, as a demonstration that they all, in a sense, resulted from just such complex activity—they are the surviving testimony. The source is an outstanding monument to the double repertory of late medieval Italian and French secular song, owing not necessarily to the "industry and versatility" of any one scribe, but rather to an artistically cosmopolitan north-Italian musical culture that was represented by—and, in turn, supported—a variety of institutions, musicians, and scribes.

Notes

1. See in particular, Kurt von Fischer, "The Manuscript Paris, Bibl., nat., nouv. acq. fr. 6771 (Codex Reina = PR)," *Musica Disciplina* 11 (1957):38–78; Nigel Wilkins, "The Codex Reina: A Revised Description," *Musica Disciplina* 17 (1963): 57–73; von Fischer, "Rely to N. E. Wilkins' Article on the Codex Reina," *Musica Disciplina* 17 (1963): 75–77; Nigel Wilkins, "A Critical Edition of the French and Italian Texts and Music Contained in the Codex Reina" (Ph.D. diss., University of Nottingham, 1964); Ursula Günther, "Bermerkungen zum älteren fran-

zösischen Repertoire des Codex Reina," *Archiv für Musikwissenschaft* 24 (1967): 237–52.

2. Principally in Wilkins, "A Critical Edition," vol. 2; idem, ed., *A 14th-Century Repertory from the Codex Reina* (American Institute of Musicology, 1966); idem, ed., *A 15th-Century Repertory from the Codex Reina* (American Institute of Musicology, 1966); Nino Pirrotta, ed., *The Music of Fourteenth-Century Italy,* 5 vols. (American Institute of Musicology, 1954–64); Willi Apel, ed., *French Secular Music of the Late Fourteenth Century* (Cambridge, 1950); idem, ed., *French Secular Compositions of the Fourteenth Century,* 3 vols. (American Institute of Musicology, 1970).

3. Research for this phase of the study was supported by a grant from the Martha Baird Rockefeller Fund for Music.

4. A more thorough discussion of the nature of readings and variants forms part of this author's doctoral dissertation at New York University, "The Transmission of Trecento Secular Polyphony: Manuscript Production and Scribal Practices in Italy ca. 1400."

5. Three vols. (Leipzig, 1904), I:260ff. See Wilkins, "A Critical Edition," pp. 1–5, for a discussion of early studies and descriptions.

6. Friedrich Ludwig, review of J. Wolf's *Geschichte,* in *Sammelbände der Internationalen Musikgesellschaft* 6 (1904–05): 616; idem, *Guillaume de Machaut: Musikalische Werke,* 4 vols. (Leipzig, 1926–54), 2:24–25.

7. Leo Schrade, ed., *The Works of G. de Machaut,* 2 vols. (Monaco, 1956), 1:48.

8. See note 1 above.

9. Von Fischer, "The Manuscript," pp. 38–44.

10. See also Ursula Günther, "Die Anwendung der Diminution in der Handschrift Chantilly 1047," *Archiv für Musikwissenschaft* 17 (1960): 6–7.

11. Von Fischer, "The Manuscript," pp. 74–77: gathering 8 (fols. 89–108), inserted folios 109 and 110, gathering 9 (fols. 111–18), and isolated folio 119. Folios 125–27 remain structurally undefined.

12. Von Fischer, "The Manuscript," p. 39.

13. Wilkins, "A Critical Edition," pp. 6–17.

14. Wilkins, "The Codex Reina," pp. 57–66.

15. Ibid., p. 60.

16. Ibid.

17. Ibid., pp. 61–66.

18. Ibid., p. 66.

19. Allan Stevenson, *The Problem of the Missale speciale* (Pittsburgh, 1967), pp. 26–27.

20. See Allan Stevenson, "Chain-Indentations in Paper as Evidence," *Studies in Bibliography* 6 (1954): 181–95. The mould sides have been noted by a small italic *m* in diagram 1. Thus the mould side of bifolium 1/12 is the side comprising folios 1r and 12v.

21. Discussed most recently in G. Thomas Tanselle, "The Bibliographical Description of Paper," *Studies in Bibliography* 24 (1971): 46–48.

22. Allan Stevenson, "Watermarks are Twins," *Studies in Bibliography,* 4 (1951–52): 57–91. See also the discussion in Stevenson, *The Missale speciale,* pp. 34–35, concerning the important distinguishing feature of sewing-dot patterns.

23. See Stevenson, *The Missale speciale,* p. 52, where he discusses the two basic sizes of paper sheets, elaborating on Briquet's distinction. By the fifteenth century, there existed two basic sizes: small paper *(forma minor)* measuring approximately 30 cm in height, and large paper *(forma major)* approximately 40

cm in height. Unless the Reina Codex paper underwent excessive trimming, it would seem to fall under the small paper size.

24. Jean Irigoin, "La datation par les filigranes du papier," *Codicologica* 5 (1980): 26–29.

25. Von Fischer, "The Manuscript," p. 39; C. M. Briquet, *Les Filigranes*, ed. Allan Stevenson (Amsterdam, 1968), between pages 8 and 9 of the original French introduction.

26. Vladimir A. Mošin and Deid M. Traljić, *Filigranes des XIIIe et XIVe SS* (Zagreb, 1957).

27. Aurelio and Augusto Zonghi, *Zonghi's Watermarks* (Hilversum, 1953).

28. For a similar but smaller mark, see Gerhard Piccard, *Die Kronen-Wasserzeichen* (Stuttgart, 1961), Abteilung I, Watermark, no. 146 [Venice, 1397–1403].

29. The two halves of each single bifolium were ruled together, the prickmarks in the margins guiding the placement of the staves. This procedure seems to have been followed for the entire manuscript, and may be assumed by the reader unless specifically indicated to the contrary.

30. Eight prickmarks, not always visible, appear to have guided the ruling of the staves, and are arranged so that each mark corresponds with the top line of a stave. As was common practice, the entire gathering was pricked at one time. Since for this manuscript the pricking resulted in irregularly spaced and misaligned prickmarks, the design formed in the margins must be identical throughout all leaves prepared together. Thus any discrepancies in the pattern of pricking should signal to the investigator the possibility of added or replaced folios.

31. These strips are found on the outside of the outer bifolium and on the inside of the inner bifolium of a gathering:

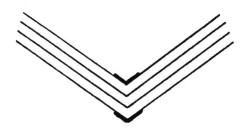

32. A narrowing of stave height can occur when the rastrum is turned to the right or left of an imaginary perpendicular axis to the staves.

33. This red guide line can often be followed visually into the fold of a bifolium and seen to continue onto the conjugate leaf.

34. In the interest of avoiding further confusion, I have used the final letters of the alphabet to distinguish the scribes in the present analysis.

35. Von Fischer, "Reply," p. 75.

36. I draw attention in particular to the presence of red capitals on fols. 77v and 80r, and the fact that *Au tornai de mors* (on fol. 77r) lacks a folio number in the index. The rondeau, *En tes doulz flans*, on fol. 77v, is missing altogether from the

index. At the bottom of fol. 67v, a different hand has added the words (catchword?) "Amor, merçe per dio." Wilkins, in "A Critical Edition," p. 392, goes against a suggestion made by von Fischer in proposing that the composition in question might be "Merci pour Dieu" in the manuscript Lucca, Archivio di Stato 184 (Mancini Codex). In neither his 1963 article nor in his dissertation does Wilkins give a full account of the occurrence of red ink in gatherings 6 and 7. It should be noted, moreover, that the unique position of paper type 4 as inner- and outermost bifolia in gatherings 6 and 7 may by itself signal a distinct layer of compilation.

37. Already noted by von Fischer, "The Manuscript," p. 46, and Günther (see note 10 above).

38. From a paleographical point of view, this is based on the belief that a scribe will reveal his own distinctive traits in the execution of the more "mechanical" aspects of his craft (clefs, sharps and flats, semiminim flags, etc.) than in the execution of willful features such as the decoration of final notes and divisional bar lines (demanding conscious attention and therefore prey to imitation or adaptation from one scribe to another).

39. This form of the sharp sign was also used by the scribe who copied a Paduan fragment (Padua, Biblioteca universitaria, MS 684), belonging with two others to a common larger manuscript known as Pad A (once part of the library of the Benedictine monastery of Santa Giustina in Padova). See Dragan Plamenac, "Another Paduan Fragment of Trecento Music," *Journal of the American Musicological Society* 8 (1955): 165–81. For an example of this type of sharp sign, see Willi Apel, *The Notation of Polyphonic Music 900–1600* 5th ed. (Cambridge, 1953), facsimile #74 (Reina Codex, fol. 17r). The drawing of dots within the sign may be associated with the Marchettan subdivision of the semitone.

40. Contrasting with some of the later, smaller hands. See, for example, the addition by Scribe S of the word *Triplum* to Scribe W's copying on fol. 58v.

41. The same form may also be found in the Paduan fragment known as Pad C (Padua, Biblioteca universitaria, MS 658). See Kurt von Fischer, "Padua und Paduanes Handschriften," *Die Musik in Geschichte und Gegenwart* (Kassel, 1949–1979), 10: 571–72. For an example of T's hand, see F. Alberto Gallo, "Ricerche sulla musica a S. Giustina di Padova all' inizio del II quattrocento; due 'siciliane' del trecento," *Annales Musicologiques* 7 (1978): 43–50 (facsimile facing p. 45 is of Reina Codex, fol. 26r); see also Wilkins, "The Codex Reina," p. 67 (fols. 27v–28), and Apel, *French Secular Music,* plate 7 (fol. 84v).

42. For an example of U's hand, see von Fischer, "The Manuscript," facsimile facing p. 48 (fol. 39v: *Donna fallante*).

43. For similar use, see scribal practice in fragment Grottaferrata, Biblioteca della Badia Greca, MS E.β.XVI (inventoried and discussed by K. von Fischer, "Eines neues Trecentofragment," *Festschrift für Walter Wiora* [Basel, 1967], pp. 264–68) and the Chantilly Codex (Gilbert Reaney, "The Manuscript Chantilly, Musée Condé 1047," *Musica Disciplina* 8 (1954): 59–113. See Carl Parrish, *The Notation of Medieval Music* (New York, 1959), plate 60 (Reina Codex, fol. 85r).

44. For a facsimile of fol. 45r, see Wilkins, "The Codex Reina," p. 70.

45. That this hand clearly differs from others in the manuscript was noted in Giuseppe Corsi, *Poesie musicali del trecento* (Bologna, 1970), p. lxxi. See Wilkins, *A Fourteenth-Century Repertory,* p. xxi (Reina Codex, fols. 56v–57).

46. For an example of Y's hand, see Wilkins, "The Codex Reina," p. 68 (fols. 63v–64).

47. The section devoted to Bartolino in the Squarcialupi Codex (Florence, Biblioteca Laurenziana, MS Palatino 87) begins, in fact, with *La douce cere* and an illuminated letter *L*. Such intended prominence would lend even more weight to the thesis that the Reina Codex originated in Padua; see Wilkins, "The Codex Reina," p. 64, and Anne Hallmark, "Some Evidence for French Influence in Northern Italy, c.1400," *Studies in the Performance of Late Mediaeval Music* (Cambridge, 1983), pp. 220–21. For a *terminus ante quem* of 1401 for part 1 of the manuscript, see Pierluigi Petrobelli, "Some dates for Bartolino da Padova," *Studies in Music History: Essays for Oliver Strunk*, ed. Harold Powers (Princeton, 1968), pp. 94–98.

48. *Sotto l'imperio* had also been chosen to head the Jacopo section in both the Squarcialupi Codex and MS Paris, Bibliothèque Nationale, fonds it. 568. In the latter, it is indeed the first composition of the entire collection. In this light, Jacopo's madrigal *Lo lume vostro* (Reina Codex, fol. 1r) can be regarded as a somewhat later addition, most probably belonging to the layer of compilation that included further Jacopo and Giovanni works (see the discussion to follow).

49. The incomplete states of fols. 13r *(Imperial sedendo)* and 18r *(Qual novità)* were not considered by either von Fischer or Wilkins in their discussions of the structure of the source. In fact, Wilkins went so far as to suggest that Bartolino's *Amor che nel pensier* (the work occupying most of fol. 18r) began with a "monodic" instrumental prelude" twelve bars in length, and he also remarked on the composition's unusual rhythmic and harmonic schemes [!]; see Wilkins, "A Critical Edition," p. 319.

50. This contratenor part is thought to have been composed by Matteo da Perugia, and appears, alone, in the manuscript Modena, Biblioteca Estense, α.M.5.24, fol. 3v (see Ursula Günther, "Das Manuskript Modena, Biblioteca Estense, α.M.5.24 [*olim* lat.568 = Mod]," *Musica Disciplina* 24 [1970]: 17–67).

51. See Nino Pirrotta, *Paolo Tenorista* (Palm Springs, 1961), p. 19. Conversely, the anonymous *Io son un pellegrin* (fols. 27v–28) gains in acceptance as a Landini composition, for it belongs to the short series of Scribe S's addition of Landini works—*Sie maledetta l'or* (fol. 28r), *Amor, donna, chi t'ama* (fols. 26v–27), and *Donna, s'i' t'o fallito* (fol. 34r).

52. The latter known also as "Arrigo" in manuscript Paris, Bibliothèque Nationale, fonds it. 568, fols. 96v–97.

53. See Nino Pirrotta, "New Glimpses of an Unwritten Tradition," in *Words and Music: The Scholar's View. A Medley of Problems and Solutions Compiled in Honor of A. Tillman Merritt* (Cambridge, Mass., 1972), pp. 271–91. In addition, see F. Alberto Gallo, "Ricerche."

54. It should be noted that Scribe X's contributions may all be compositions by Bartolino. The unattributed *Fa se'l bon servo* (fol. 40v) and, particularly, *La nobil scala* (fols. 40v–41) have been ascribed with some likelihood to Bartolino by a number of investigators, including Wilkins, "The Codex Reina," p. 64.

55. Already noted by Pirrotta, "New Glimpses," p. 274.

56. Part 2 also subdivides into "layers," with fols. 53r–62v forming one section, and gatherings 6 and 7, another. This is signaled, among other things, by changes in the foliation at fols. 51r and 62v and by the dimensions of the writing space.

57. See note 36 above.

58. Briefly mentioned in this author's "The Structure of MS Panciatichi 26 and the Transmission of Trecento Polyphony," *Journal of the American Musicological*

EXAMPLE 1. *Je voy le bon tens venir* (Reina Codex, fol. 85v).

Society 34 (1981): 418 n.29. Subsequent to this author's independent discovery of the identity and nature of the arrangement on fol. 85v of the Reina Codex, the following sources were made available: Robert L. Huestis, "Scribal Errors in the Faenza Codex: A Clue to Performance Practice?" *Studies in Music* [University of Western Australia] 10 (1976): 52–61, and idem, "Contrafacta, Parodies, and Instrumental Arrangements from the Ars Nova," (Ph.D. diss., University of California at Los Angeles, 1973), 1:134 and 138–39. In both of these sources, Professor Huestis had correctly identified the vocal model of the arrangement on folio 85v of the Reina Codex; I do not agree, however, with his assertion that both the model and arrangement were copied by the same scribe. See example 1 for a modern edition of this arrangement.

59. Leo Schrade, ed., *The Works of Francesco Landini*, Polyphonic Music of the Fourteenth Century, IV (Monaco, 1958–59), commentary, pp. 80, 89, and 102–3.

60. Schrade, *Landini*, commentary, p. 80.

61. The other sources are MS Paris, Bibliothèque Nationale, fonds it. 568, fols. 104v–105; Squarcialupi Codex, fol. 150v.

62. See note 58 above. The scribe abandoned his arrangement for lack of a proper notational solution in his score of the syncopated tenor passage in the vocal model. For a proper contemporary method of notating syncopations in a score format—through the omission of bar lines—see, for example, the Faenza Codex arrangement of Antonio Zacara da Teramo's *Un fior gentil* (fol. 82r), a modern edition of which is to be found in Dragan Plamenac, *Keyboard Music of the Late Middle Ages in Codex Faenza 117* (American Institute of Musicology, 1972), pp. 106–7, bars 27–28 and 38–39.

63. See Margaret Bent, "Some Criteria for Establishing Relationships Between Sources of Late-Medieval Polyphony," *Music in Medieval and Early Modern Europe: Patronage and Sources* (Cambridge, 1981), pp. 309–10.

64. See von Fischer, "The Manuscript," p. 57 of the inventory. It may be of significance that the contratenor part of *Alba colunba* varies markedly from that in the Squarcialupi Codex; perhaps this is further indication that it may have been added to an original two-part composition. (In two examples, *Imperialle sedendo* [fols. 22v–23] and *Non correr troppo* [fol. 22r], Scribe S transmitted Bartolino works as fully texted, two-part compositions when they appear as three-part works elsewhere.) Scribe S, in fact, can be associated with fully texted texture, and may be suspected of adding words to parts that were textless in his exemplars: note the heavily ligated—yet texted—tenors in Landini's *Per seguir la sperança* (fol. 48, most likely copied by Scribe S) and *Chi preghio vole* (fol. 51).

65. Scribe X's transmission of such rhythmic refinements in *I bei senbianti* (fols. 45v–46) may be seen to contrast with Scribe T's presumed reluctance to do so in *Sempre, dona, t'amay* (fol. 15v).

66. For an excellent discussion of the problems associated with notated syncopations, see Michael P. Long, "Musical Tastes in Fourteenth-Century Italy: Notational Styles, Scholarly Traditions, and Historical Circumstances" (Ph.D. diss., Princeton University, 1981), pp. 100–102.

67. See Corsi, *Poesie*, introduction, pp. lxxxi–xciv, and Pierluigi Petrobelli, " 'Un leggiadretto velo' ed altre cose petrarchesche," *Rivista Italiana di Musicologia* 10 (1975): 32–45.

68. Wilkins considered the texts of this entire section of the manuscript (gatherings 6 and 7) to be corrupt; see Wilkins, "A Critical Edition," p. 9, and idem, ed., *A 14th-Century Repertory*, critical notes passim.

69. Bent, "Some Criteria," pp. 297–313.

5

A Checklist of Books and Articles Containing Reproductions of Watermarks

Phillip Pulsiano

The following bibliography lists books and articles containing reproductions of watermarks, and is intended as a preliminary guide for researchers interested in compiling and comparing watermark reproductions. The checklist updates Allan Stevenson's bibliography in *The New Briquet* (Amsterdam: The Paper Publications Society, 1976), yet is not as selective. Instead, I have tried to provide users with a wide range of sources in which reproductions of watermarks can be found. Thus, a reader interested in the use of watermarks as forensic evidence, in the recent applications of watermark research in musicology, in the watermarks of various paper mills, or in the aesthetics of watermarks, will find such items listed here.

This checklist also supplements, to a certain extent, Irving P. Leif's *An International Sourcebook of Paper History* (Hamden, Conn.: Archon Books, 1978), Kate Frost's "Supplement to Leif: A Checklist of Watermark History, Production and Research" (*The Direction Line* 8 (1979): 33–56), and Jack Weiner and Kathleen Mirkes's much neglected *Watermarking* (Appleton, Wis.: The Institute of Paper Chemistry, Bibliographic Series, no. 257, 1972). Many potential sources of watermark reproductions found in journals such as *Der Büromarkt* (Aachen), *Livéltárí Hiradó,* and *Basler Nachrichten* could not be included here, since it has been my policy to see the source wherever possible, or at least to find secondary verification of watermark reproductions.

The checklist is alphabetically arranged. The number in parentheses following each entry indicates the number of watermark reproductions contained in each work. It was my original intention to include entries through 1981. Since then, however, a number of important works have appeared, most notably, G. Piccard's additions to the *Findbuch* series and T. Gravell and G. Miller's catalogue of American watermarks. I have thus revised this checklist to include more recent publications, albeit selectively.

I wish to thank Donna Sammis and Robert Lobou of the Frank J. Melville Memorial Library at the State University of New York at Stony Brook for their assistance in locating many of these items. I am also grateful to Professor Kate Frost of The University of Texas at Austin for kindly supplying me with references from her own unpublished watermark research. Above all, I wish to thank Professor John Bidwell for reading through this checklist in its early stages. His comments and suggestions have proven invaluable in the preparation of this work.

A

1. Abrams, T. M. "The History and Artistry of Watermarks." *Pulp and Paper Magazine of Canada* 64, no. 11 (November 1963): 79–81. (4)
2. Alberti, Karl. "Die ehemaligen Papiermühlen im Ascher Gebiete und ihre Papier-Wasserzeichen." *Unser Egerland* 31 (1927): 69–77. (9)
3. Alibaux, Henri. *Les premières papeteries françaises*. Paris, 1926. (27)
4. Ames, Joseph. *Typographical Antiquities*. London, 1749. Revised edition by Thomas Frognall Dibdin. London, 1810–19. Reprint. Hildesheim, 1969. (38)
5. *Angelelli, Onofrio. L'industria della carta e la famiglia Miliani in Fabriano. Fabriano, 1930.* (22)
6. Antonelli, D. G. *Ricerche bibliografiche sulle Edizioni Ferraresi del secolo XV*. Ferrara, 1830. (23)
7. Asenjo, J. L. "'Arrigorriaga' Watermarks Up to 1936." *Investigación y Tecnica del Papel* 3, no. 10 (1966): 849–89. (65)
8. Ataíde e Melo, A. F. de. *O papel como elemento de identificação*. Lisbon, 1926. (213)
9. Audin, M. "Le Centre papetier Ambert-Beaujeu-Annonay." *Contribution à l'Histoire de la Papeterie en France* 9 (1943): 19–74. (6)
10. Bacîru, L. "Valoarea documentară a filigranelor, cu privire

specială asupra cărtilor romaneşti tipărite in secolul al XVI-lea" (The Documentary Value of Watermarks With Special Reference to Books Printed in Rumania in the 16th Century). *Studii şi cercetări de documentare şi bibliologie* (Bucharest) 7 (1965): 273–98. (54)

B

11. Badecki, K. *Znaki wodne w księgach archiwum miasta Lwowa, 1382–1600 r.* (Watermarks in Volumes in the Lvov Municipal Archives) Lwów, 1928. (166)
12. Bailo, L. *Sulle prime cartiere in Treviso.* Treviso: 1887. (32)
13. Balston, Thomas. *William Balston, Paper Maker, 1759–1849.* London, 1954: Reprint. London and New York, 1979. (2; includes 12 countermarks)
14. ———. *James Whatman, Father & Son.* London, 1957.
15. Barlow, Thomas D. *Albert Dürer: His Life and Work.* London, 1925. (Originally a lecture delivered to The Print Collector's Club, 15 November 1922.) (18)
16. Barone, Nicola. "Le filigrane delle antiche cartiere ne'documenti dell'Archivio di Stato in Napoli dal XIII al XV secolo." *Archivio Storico per le Province Napoletane* (Naples) 14 (1889): 69–96. (72)
17. Basanoff, A. "L'Emploi du papier à l'Université de Paris (1430–1473)." *Bibliothèque d'humanisme et renaissance* (Geneva) 26 (1964): 305–25. (26)
18. ———. *Itinerario della carta dall'Oriente all'Occidente e sua diffusione in Europa.* Milano, 1965. (119)
19. Basanta Campos, J. L. "Watermarks in Galician Documents [at] the Provincial Historical Archives of Pantevedra." *Investigación y Tecnica del Papel* 4, no. 14 (1967): 877–93. (100)
20. Bauer, E. *Histoire des papeteries de Serrières, 1477–1934.* Zofingue, 1934. (22)
21. Bayley, Harold. "Notes on Watermarks." *Booklover's Magazine* 6 (1906): 65–71. (60)
22. ———. "Alchemy and the Holy Grail." *Baconiana*, 3d ser., 5 (1907): 28–48. (Lecture delivered before the Bacon Society) (22)
23. ———. "The Alchemyst At-one-ment." *Baconiana*, 3d ser., 5 (1907): 111–15. (10)
24. ———. "The Romance of Papermarks." *The Bibliophile* 1 (1908): 93–6. (13)

25. ———. *A New Light on the Renaissance, Displayed in Contemporary Emblems.* London and New York, 1909: Reprint. 1967. (409)

26. ———. *The Lost Language of Symbolism; an Inquiry into the Origin of Certain Letters, Words, Names, Fairy-Tales, Folk-Lore, and Mythologies.* London and New York, 1912: Reprint. Philadelphia, 1913; New York, 1951; London, New York, and New Jersey, 1952, 1974; London and New York, 1957, 1968. First published in two volumes. (c. 1500)

27. Beadle, Clayton. "Die Herstellung von Wasserzeichen, ihre Entwicklung in Hand- und Maschinenpapieren." *Der Papier-Fabrikant* 38 (1906): 2055–58; 39:2111–15; 41:2224–26; 42:2277–78. (27)

28. ———. "The Development of Watermarking in Hand-Made and Machine-Made Papers." *Journal of the Society of the Arts* 54 (11 May 1906): item no. 2791. (27)

29. ———. "The Cl. Beadle Collection of Watermarked Papers." *Pulp and Paper Magazine of Canada* 6 (1908): 77–80.

30. ———. "The Study of Ancient Watermarks." *The World's Paper Trade Review* 49, no. 14 (April 1908): 569–73. (12)

31. ———. "A Collection of Historical Watermarked Papers." *Pulp and Paper* (London) 28 (April 1908): 155–58.

32. ———. "The Development of Watermarking in Paper." *Paper* 3, no. 3 (5 April 1911): 9–12, 28, 30, 32. (28)

33. Beans, G. H. *Some Sixteenth Century Watermarks Found in Maps Prevalent in Italian Assembled-to-Order (IATO) Atlases.* Jenkintown, Penn., 1938. (76)

34. Blake, N. F. *Caxton: England's First Publisher.* London, 1976. (8)

35. Blaser, Fritz. "Luzerner Buchdrukerlexikon. I Teil: Umfassend die Zeit von der Einführung der Buchdruckerkunst bis zum Jahre 1798." *Der Geschichtsfreund* 84 (1929): 142–72. (7)

36. Blücher, G. "Filigranele braşovene şi tipărturile chirilice dîn secolul al XVI-lea" (Braşov [Kronstadt] Watermarks and 16th Century Cyrillic Printed Books). *Revista Bibliotecilor* (Bucharest) 20 (1967): 421–26. (9)

37. Bockwitz, H. H. "Zum 50. Todestage von Friedrich Keinz, 1901. Ein bayrischer Wasserseichenforscher und Verehrer von Ch. M. Briquet." In *The Briquet Album*, 1952. (1)

38. ———. "Eine alt-holländische Papier-Wassermühle." *Der Altenburger Papierer* 10 (1936): 552–54: Reprinted in *Archiv für Buchgewerbe und Gebrauchsgraphik* 73 (1936): 131–36. (1)

39. Bodemann, E. *Xylographische und typographische Inkunabeln der Königlichen öffentlichen Bibliothek zu Hannover.* Hannover, 1866. (200)
40. Boesch, Hans. "Die Hohensollern und die Papierfabrikation in Franken." *Papier-Zeitung* 22 (1897). (10)
41. Bofarull y Sans, Fr. de A. de. *La heráldica en la filigrana del papel.* Barcelona, 1901. (129)
42. ———. *Los animales en las marcas del papel.* Villanueva y Geltrú, Barcelona, 1910. (762)
43. ———. *Heraldic Watermarks, or, La heráldica en la filigrana del papel.* (Supplement to Monumenta Chartæ Papyraceæ Historiam Illustrantia.) English translation by A. J. Henschel. Hilversum, 1956. (129)
44. Bogdán, István. "Vizjelek és vizjelkutatás" (Watermarks and Watermark Studies). *Papir- es Nyomdatechnika* (Budapest) 8 (1956): 25–28. (11)
45. ———. "A vizjelkutatás problémái (vizjeguijtésünk modszertana)." (Problems of Watermark Research [Methods of Collection and Arrangement]). *Levéltári kozlemények* (Budapest) 30 (1959): 89–108. (4)
46. ———. "A Lékai (Hámori) papírmalom a XVIII. században" (The Papermill of Lockenhaus [Hammer] in the 18th Century). In *Történelmi szemli*, pp. 46–93: Budapest, 1960. German summary, pp. 92 ff. (13)
47. ———. *A magyarországi papíripar története, 1530–1900.* Budapest, 1963. *(History of Papermaking in Hungary.)* (37)
48. ———. "A 'vizjelirás' fejlödése" (The Development of "Watermark Script"). *Papiripar* 8 (1964): 170–78. (21)
49. ———. "Watermarks as Trademarks." *Papiripar* 14, no. 2 (1970): 64–70.
50. Bordeaux, Raymond. "Prix de divers objets de librairie fournis à une église de compagne à la fin du XVIe siècle." *Bulletin du Bouquiniste* 37 (1858): 328–29. (1)
51. Bošnjak, M., V. Hofman, and V. Putanec. "Vodeni znakovi hrvatskih inkunabula" (Watermarks in Croat Incunabula). *Bulletin Zavoda za likovne umjetnosti JAZU* 11, no. 3 (1963): 20–50. (25)
52. Boudon, G. *Notes sur quelques filigranes de papiers des XIVe et XVe siècles et de la première moitié du XVIe.* Amiens, 1889. (37)
53. Boyer, H. and Vallet de Viriville. *Filigranes de papier du 15e siècle aux armes des familles Coeur et de Bastard.* Paris, 1860.
54. Bradley, W. A. "Significance of Early Watermarks." *Printing*

Art 16 (1910): 176–80. (4)

55. Brandenburg, Sieghard. "Bemerkungen zu Beethovens Op. 96." *Beethoven-Jahrbuch* 9 (1977): 11–25.

56. ———. "Ein Skizzenbuch Beethovens aus dem Jahre 1812: zur Chronologie der Petterschen Skizzenbuches." In *Zu Beethoven*, edited by Harry Goldschmidt, pp. 117–48. Berlin, 1979.

57. Brătulescu, Victor. *Miniaturi şi manuscrise din Museul de Artă Religioasă*. Bucharest, 1939. (26)

58. Briquet, Charles M. "Notices historiques sur les plus anciennes papeteries suisses." Lausanne: L'Union de la Papeterie, 1883–1885. (30)

59. ———. "Recherches sur les premiers papiers employés en Occident et en Orient du Xe au XIVe siècle." *Mémoires de la Société Nationale des Antiquaires de France*, 5th ser., 16 (1885): 133–205. (17)

60. ———. "Papiers et filigranes des Archives de Gênes, 1154 à 1700" (1887). In *Briquet's Opuscula* (1955), pp. 171–220: Book form: *Papier et filigranes de Gênes*. Genève, 1888. (594)

61. ———. "Lettre à M. le chevalier I. Giorgi préfet de la Bibliothèque de Palerme, sur les papiers usités en Sicile à l'occasion de deux manuscrits en papier de coton." *Archivio Storico Siciliano*, (n.s.) 17 (1892): Reprinted in *Briquet's Opuscula* (1955), pp. 228–34. (41)

62. ———. "Les anciennes papeteries du duché de Bar, et quelques filigranes barrois de la second moitié du XVe siècle." *Le Bibliographe Moderne* 1 (1898): Reprinted in *Briquet's Opuscula* (1955). (8)

63. ———. "La Date de trois impressions précisée par leurs filigranes." *Le Bibliographe Moderne* 2 (1900): 113–33. (5)

64. ———. "La Papeterie sur le Rhône à Genève et les papiers filigranés à l'écu de Genève." In *Nos anciens et leurs oeuvres*. Genève, 1901. Reprinted in *Briquet's Opuscula* (1955), pp. 299–309. (27)

65. ———. "Notions pratiques sur le papier." *Le Bibliographe Moderne*, nos. 1–2 (1905): 5–36. (26)

66. ———. *Les Filigranes: Dictionnaire historique des marques du papier dès leur apparition vers 1282 jusqu'en 1600, avec 39 figures dans le texte et 16, 112 facsimilés de filigranes*. Genève, 1907. 2d ed. 4 vols. with supplementary text. Leipzig, 1923; Reprint. Amsterdam, 1968 (see *The New Briquet*); New York, 1969, 1977; Hildesheim, 1977.

67. ———. *The Briquet Album.* See Labarre.
68. ———. *Briquet's Opuscula: The Complete Works of Dr. C. M. Briquet Without Les Filigranes.* (Monumenta Chartæ Papyraceæ Historiam Illustrantia, IV). Hilversum, 1955. (683)
69. ———. *The New Briquet . . . Les Filigranes. A Facsimile of the 1907 Edition With Supplementary Material Contributed by a Number of Scholars.* Edited by Allan Stevenson. 4 vols. Amsterdam, 1968.
70. Buchmann, G. *Geschichte der Papiermacher zu Oberweimar. Neue Beiträge zur Geschichte der Stadt Weimar.* Weimar, 1936. (16)
71. Budka, W. "Filigrany z herbami Łodzia i Lis" (Watermarks with the Lodzia and Lis Arms). *Silva rerum* (Cracow) 4 (1918): 180–82. (2)
72. ———. "Papiernia w Odrzykoniu i Mniszku" (The Papermills at Odrzykoń and Mniszek). *Przegląd biblioteczny* (Cracow) 5 (1931): 61–66. (5)
73. ———. "Papiernia w Balicach" (The Papermill at Balice). *Archeion* (Warsaw) 13 (1935): 30–49. (10)
74. ———. "Herby Prus Królewskich i Prus Książęcych jako znaki wodne" (Arms of East and West Prussia as Watermarks). *Przegląd biblioteczny* (Cracow) 11 (1937): 289–92. (3)
75. ———. "Papiernia w Mniszku" (The Mniszek Papermill). *Przegląd Papierniczy* (Łódź) 7 (1951): 121–25. (7)
76. ———. "Papiernie poznańskie" (Poznań Papermills). *Przegląd Papierniczy* 10 (1954), 216–21: 251–53. (13)
77. ———. "Papiernie w Lubinie i Kocku" (Papermills at Lublin and Kock). *Archeion* 25 (1956): 257–75. (6)
78. ———. "Paper Mills in Mostki, Drugnia, and Slopiec." *Przegląd Papier* 26, no. 8 (1970): 278–80.
79. Budka, W., ed. *Papiernie w Polsce XVI wieku: prace Franciszka Piekosińkiego, Jana Ptaśnika, Kazimierza Piekarskiego* (Papermills in 16th Century Poland: Works by . . .). Ossolineum, 1971. (276)

C

80. Calflisch, E., and L. Calflisch. *Aus der Geschichte der Zürcher Papiermühle auf dem Werd, 1471–1700.* Zürich, 1963. (2)
81. Camus, A. G. *Notice d'un livre imprimé à Bamberg en 1462.* Paris, 1799. (3)

82. Chroust, Anton. *Monumenta palaeographica*. München, 1902–17. (2)
83. Churchill, W. A. *Watermarks in Paper in Holland, England, France, etc. in the XVII and XVIII Centuries and Their Interconnection*. Amsterdam, 1935. Reprint. 1967. (578)
84. Clapperton, R. H. *Paper: An Historical Account of Its Making by Hand from the Earliest Times Down to the Present Day*. Oxford, 1934. (26)
85. Claudin, A. *Origines de l'imprimerie à Albi en Languedoc, 1480–84*. Paris, 1880. (5)
86. Clausen, Hans Dieter. *Händels Direktionspartituren ("Handexemplare")*. Hamburger Beiträge zur Musikwissenschaft, Band 7. Hamburg, 1972.
87. Clauss, Friedrich. *Memminger Chronik*. Memmingen, 1894.
88. Clemensson, G., ed. *Klippans Pappersbruk, 1573–1923*. Lund, 1923. (169)
89. Clemensson, G., ed. *En Bok om Papper*. Uppsala, 1944. (47)
90. Cohendy, Michel. "Notes sur la papeterie d'Auvergne antérieurement à 1790 et les marques de fabrique des papeteries de la ville et baronnie d'Aubert et ses environs." *Mémoires de l'Académie de Science, Belle-Lettres et Arts de Clermont* 4 (1862): 196–217. (41)
91. Collijn, J. *Katalog der Incunabeln der Königlichen Bibliothek in Stockholm*. Stockholm, 1914.
92. Corraze, R. "L'Industrie du papier à Toulouse (1500–1530)." *Contribution à l'Histoire de la Papeterie en France* 2 (1935): 94–104. (14) Supplement in ibid. 6 (1941): 45–46.
93. ———. "Un moulin à papier à Toulouse au commencement du 15e siècle (1419)." *Contribution à l'Histoire de la Papeterie en France* 6 (1941): 13–19. (4)

D

94. Dangel, A. "Von einer bisher unbekannten Papiermühle zu Schwäbisch Gmünd." *Papiergeschichte* 8 (1958): 61–62. (3)
95. Decker, V. *Náčrt dejín ručného papiernictva na Slovensku* (A Brief History of Hand Papermaking in Slovakia). Martin, 1956. (92)
96. ———. *Priesvitky archívu mesta Kremnice v zbierke Pavla Križku* (Watermarks from the Kremnica Archive in the Collection of Pavol Križko). Martin, 1956. (235)

97. ———. "Watermarks of Kremnica Mills." *Papír a celulosa* 17, no. 11 (1962): 255–57.
98. ———. "The Paper Mill in Ochtina." *Papír a celulosa* 18, no. 11 (1963): 225–27.
99. ———. "Vznik a vývoj papiernictva na Slovensku v 16. storočí a v prvej polovici 17. storočia" (The Rise and Development of Papermaking in Slovakia in the 16th and 17th Centuries). *Sborník Národního muzea v Praze* (Prague), ser. 100, no. 11 (1966): 147–60. (10)
100. Degaast, Georges. "Les Vieux moulins à papier d'Auvergne," *Gutenberg-Jahrbuch*, 1963, pp. 9–13.
101. De la Serna Santander. *De la Serna Santander catalogue des livres de sa bibliothèque*. Bruxelles, 1803. (147)
102. Demény, K. "Papiergeschichte des XVI. Jahrhunderts in den Forschungen der rumänischen Historiker (Probleme und Zukunftsaussichten)." *International Congress of Paper Historians, Communications* 7 (1967): 17–32. (6)
103. Denne, Samuel. "Observations on Paper-Marks." *Archaeologia* 12 (1795): 114–31. (47)
104. Desbarreaux-Bernard T.P. *Catalogue des incunabules de la Bibliothèque de Toulouse*. Toulouse, 1878. (294)
105. Dîmboiu, A. *De la piatră la hîrtie* (From Stone to Paper). Bucharest, 1964. (26)
106. Droz, Eugénie, and C. Dalbanne. "Le Miroir de mort de Georges Chastellian." *Gutenberg-Jahrbuch*, 1928, pp. 89–92. (4)
107. "Der Dürener Raum und die Papierindustrie." *Allgemeine Papier-Rundshau* 10 (1953): 406–8, 410–12. (5)

E

108. Ebbinghaus, Paul. "Die Geschichte der Papiermühle Tycho Brahes auf der Insel. Hven." *Der Papier-Fabrikant*, Festheft, 1911, pp. 64–68. (2)
109. Ebenhöch, H. "Geschichte der Papiermühle zu Niedergösgen." *Jahrbuch für Solothurnische Geschichte* 23 (1950): 115–42. (6)
110. Eberlein, A. "Papier, Papiermacher, Papiermühlen in Mecklenburg." *Wissenschaftliche Zeitschrift der Universität Rostock* 7 (1957/58), *Gesellschafts- und Sprachwissenschaftliche Reihe*, 1:133–47. (2)

111. *100 Jahre im Dienste der deutschen Papiermacherei, 1808–1908. Geschichte und Entwicklung der greizer Papierfabrile im Besitze der Familie Günther.* Greiz, 1908. (13)

112. Eineder, G. *The Ancient Paper-Mills of the Former Austro-Hungarian Empire and Their Watermarks.* Edited and translated by E. J. Labarre. Hilversum, 1960. (Monumenta Chartæ Papyraceæ Historiam Illustrantia, VIII). (1,871)

113. Elliot, Harrison. "The Romantic Phase of Watermarks in Paper With Something of Their Significance." *Paper Maker* 17, no. 2 (1948): 11–15.

114. ———. "Connecticut's First Papermaker." *Paper Maker* 19, no. 2 (1950): 41–43.

115. ———. "The Oldest Paper Mill in the Western World and Its Historical Background." *Paper Maker* 21, no. 1 (1952): 39–46.

116. Erastov, D. P. "The Leningrad Method of Watermark Reproduction." *Book Collector* 10 (1961): 329–30. (2)

117. Ersoy, O. *XVIII. ve XIX. yüzyillarda Türkiye'de kâğit.* (Ankara Üniversitesi. Dil ve Tarih-Coğrafya Fakültesi yayimlari, 145). Ankara, 1963. (315)

118. ———. "Bursa'da kâğit fabrikasi meselesi (XV–XVI. yüzyil)." *Ankara Üniversitesi. Dil ve Tarih-Coğrafya Fakültesi dergisi,* 22 (1964): 101–15. (27)

F

119. Faloci-Pulignani, M. "Le antiche cartieri di Foligno." *La Bibliofilia* (Florence) 11 (1909): 102–27. (25)

120. Fenn, Sir John. *Paston Letters. Original Letters, Written During the Reigns of Henry VI., Edward IV., and Richard III. by Various Persons of Rank or Consequence; Containing Many Curious Anecdotes, Relative to that . . . Period of Our History . . . with Notes, Historical and Explanatory; and Authenticated by Engravings of Autographs, Facsimiles, Paper-Marks, and Seals.* 5 vols. London, 1787–1823. (99)

121. Fink, Wilhelm. "Die Deggendorfer Papiermühle." *Durch Gau und Wald* 4 (1941): 13–15. (1)

122. Fischer, G. *Beschriebung einiger typographischer Seltenheiten nebst Beitragen zur Erfindungsgeschichte der Buchdruckerkunst.* Nürnberg, 1800. (30)

123. Fiskaa, Haakon M. *Norske papirmøller og deres vannmerker, 1695–1870.* Oslo, 1973. (295)

124. Fiskaa, Haakan M., and O. K. Nordstrand. *Paper and Watermarks in Norway and Denmark*. (Monumenta Chartæ Papyraceæ Historiam Illustrantia, XIV) Translated and edited by B. J. van Ginneken-van de Kasteele, E. G. Loeber, and J. S. G. Simmons. Amsterdam, 1978. Part 1 is an edited translation of Fiskaa's *Norske papirmøller og deres vannmerker, 1695–1870*. Part 2, *Paper and Watermarks in Denmark, 1573–1899*, by O. K. Nordstrand, is translated from the Danish.

125. Fluri, Ad. "Die Papiermühle zu Thal bei Bern und ihre Wasserzeichen, 1466–1621." *Berner Taschenbuch*, 1896. (65)

126. Fry, Francis. *The Bible of Coverdale, MDXXXV. Remarks on the Titles; the Year of Publication; the Preliminary, the Water Marks, &c., With Fac-similes*. London, 1867. (10)

G

127. Gachet, Henri. "Des premiers papiers aux premiers filigranes." *Courrier Graphique* 72 (1954): 27–36. (3)

128. Gansen, P. "Die Papiermühle in Siegburg, 1490 bis 1860." *Heimatblätter des Siegkreises* 17 (1941): 21–31. (20)

129. Gaskell, Philip. *A New Introduction to Bibliography*. New York and Oxford, 1972. (18)

130. Gasparinetti, A. F. "Carta, cartiere e cartai fabrianesi." *Il Risorgimento Grafico* 25 (1938): 373–431.

131. ———. "An Old Requisition on Watermarked Paper." *Industria della Carta* 6, no. 1 (1940): 9–13.

132. ———. "Über die 'Entstellung' (Bedeutung- und Formwandel) von Wasserzeichen." *Papiergeschichte* 2, no. 3 (1952): 33–36. (6)

133. ———. "The Watermarked Names of XIV Century Italian Papermakers." *Paper Maker* 25, no. 2 (1956): 15–26. (38)

134. ———. "Ein altes Statut von Bologna über die Herstellung und den Handel von Papier." *Papiergeschichte* 6 (1956): 45–47.

135. ———. "Eine Bestellung von Wasserzeichen papier in alter Zeit." *Papiergeschichte* 8 (1958): 40–43.

136. ———. "A Curiosity of Papermaking History." *Paper Maker* 33 (1964): 33–38. (13)

137. Gauthier, J. *L'Industrie du papier dans les hautes vallées franc-comtoises du XVe au XVIIIe siècles*. Montbéliard, 1897. (62)

138. Gayoso Carreiria, G. "Notes on the Papermaking History of

the Ancient Kingdom of Valencia." *Investigación y Tecnica del Papel* 7, no. 26 (1970): 1059–91.

139. Gębarowicz, M. "Z dziejów papiernictwa XVI–XVIII w" (From the History of Papermaking in the 16th–18th Century). *Roczniki biblioteczne* (Wrocław-Warsaw) 10 (1966): 1–83. (134)

140. Geraklitov, A.A. *Filigrany XVII veka na bumage rukopisnykh i pechatnykh, dokumentov russkogo proiskhozhdeniya* (17th Century Watermarks in the Paper of Manuscript and Printed Sources of Russian Origin). Moscow-Leningrad, 1963. (1,518)

141. Gerardy, Th. "Zur Datierung des mit Gutenbergs kleiner Psaltertype gedruckten *Missale speciale.*" *Archiv für Geschichte des Buchwesens* 5 (1962–63), cols. 399–415. (22)

142. ———. *Datieren mit Hilfe von Wasserzeichen: Beispielhaft dargestellt an der Gesamtproduktion der schaumburgischen Papiermühle Arensburg von 1604–1650.* Bückeburg, 1964. (263)

143. ———. "Die Wasserzeichensammlung: der Aufbau einer Wasserzeichensammlung." *Papiergeschichte* 15, nos. 1–2 (1965): 7–14. (17)

144. Gerolt, V. "Staré opavské papírny" (The Old Opava Papermills). *Papír a celulosa* 12 (1957): 282–84. (2)

145. ———. "Filigrány staré opavské papírny" (Watermarks of the Old Opava Papermills). *Papír a celulosa* 13 (1958): 181–83. (21)

146. ———. "Papírna ve Velkých Losinach" (The Papermill at Velké Losiny). *Papír a celulosa* 16 (1961): 158–60. (7)

147. Giuliari, G. Carlo. *Della tipografia veronese.* Verona, 1871. (5)

148. Godenne, W. "Le Papier des comptes communaux de Malines datant du moyen âge." *Bulletin du Cercle Archéologique, Littéraire et Artistique de Malines* 64 (1960): 36–53.

149. Golleb Ed. "Verzeichnis der griechischen Handschriften in Oesterreich ausserhalb Weins." In *Sitzungsberichte der Philosophisch-historischen Class der Kais. Akademie der Wissenschaften,* 146. Vienna, 1903. (60)

150. Gras, L. P. *Filigranes recueillis dans quelques anciens terriers de Forez.* Ste.-Étienne, 1873. (27)

151. Gravell, Thomas L. "A New Method of Reproducing Watermarks for Study." *Restaurator* 2 (1975): 95–104. (3)

152. ———. "Watermarks: What They Are and How They Can Be Used." *Manuscripts* 32, no. 1 (1980): 4–10.

153. Gravell, Thomas L., and George Miller. *A Catalogue of Amer-*

ican Watermarks, 1690–1835. New York and London, 1979. (734)

154. ——— and George Miller. *A Catalogue of Foreign Watermarks Found on Paper Used in America, 1700–1835.* New York, 1983.

154A. Greg, W. W. "On Certain False Dates in Shakespearian Quartos." *Library,* 2d Ser., 9 (1908): 113–31, 381–409. (27)

155. Grether, Ernst. "Die Markgräfler Papiermühlen und ihre Wasserzeichen." *Das Markgräflerland* 8, no. 1 (1937): 1–11. (19)

156. Grosse-Stoltenberg, Robert. "Wasserzeichen in alten Landkarten." *Papiergeschichte* 11, nos. 5–6 (1961): 93–96. (20)

152. ———. "Stege und Stegdarstellungen." *Papiergeschichte* 13 (1963): 35–36.

158. ———. "Die Hanauer Sparren als Wasserzeichen." *Hanauer Geschichtsblätter* 21 (1966): 57–78. (65)

159. Grossman, K. "Die ältesten Wasserzeichen der Papiermühle Vlotho." *Papiergeschichte* 16 (1966): 17–18. (17)

160. Gruel, Léon. *Recherches sur les origines des marques anciennes qui se recontrent dans l'art et dans l'industrie du XVe au XIXe siècle, par rapport au chiffre quarte.* Paris and Bruxelles, 1926. (91)

161. Guntermann, Friedrich. "Die älteste Geschichte der Fabrikation des Linnenpapiers." *Serapeum,* nos. 17–18 (1845). (64)

H

162. Hassler, K. *Vortrag über die älteste Geschichte der Fabrikation des Linnenpapiers.* Ulm, 1844. (8)

163. Hausmann, Bernhard. *Albrecht Dürers Kupferstiche, Radierungen, Holzschnitte und Zeichnungen, unter besonderer Berücksichtigung der dazu verwandten Papiere und deren Wasserzeichen.* Hannover, 1861. Reprint. Würzburg, 1922. (57)

164. Hazen, Allen T. "Eustace Burnaby's Manufacture of White Paper in England." *The Papers of the Bibliographical Society of America* 48 (1954): 315–33. (4)

165. Heawood, Edward. "Sources of Early English Paper-Supply." *Library,* 4th ser., 10 (1929): 282–307. (88)

166. ———. "Sources of Early English Paper-Supply (II)." *Library,* 4th ser., 10 (1930): 427–54. (117)

167. ———. "Papers Used in England After 1600 (I)." *Library,* 4th ser., 11 (1931): 263–99. (98)

168. ———. "Papers Used in England After 1600 (II)." *Library,* 4th ser., 11 (1931): 466–98. (97)

169. ———. "Further Notes on Paper Used in England After 1600 (III)." *Library,* 5th ser., 2 (1948): 119–49. (167)

170. ———. *Watermarks, Mainly of the 17th and 18th Centuries.* Hilversum, 1950. Reprint. 1957; photo offset, 1969. (Monumenta Chartæ Papyraceæ Historiam Illustrantia, I). (4,078)

171. ———. *Historical Review of Watermarks.* Amsterdam, 1950. (223)

172. Hegg, Peter. "Ein Unikum?" *Schweizerisches Gutenberg-Museum* 40 (1954): 3–8. (5)

173. Heitz, Paul. *Les Filigranes des papiers contenus dans les archives de la ville de Strasbourg.* Strasbourg, 1902. (386)

174. ———. *Les Filigranes des papiers contenus dans les incunables strasbourgeois de la Bibliothèque Impèriale de Strasbourg.* Strasbourg, 1903. (1,328)

175. ———. *Les Filigranes avec la Crosse de Bâle.* Strasbourg, 1904. (297)

176. Helleiner, K. "Die Änfange der Papierindustrie im Viertel ob dem Wienerwald." *Jahrbuch für Landeskunde von Niederösterreich* (Vienna), n.F., 25 (1932): 175–99. (29)

177. Hellinga, Lotte, and Hilton Kelliher. "The Malory Manuscript." *British Library Journal* 3 (1977): 91–113. (3)

178. Henderson, Catherine. "A New Aid in Detecting Sophistication in Rare Book Leaves." *Direction Line* 7 (1978): 32–36. (2)

179. Henry, Avril K. "Two English Printed Books Rediscovered: Supplement to Bosanquet's *English Printed Almanaks.*" *Library,* 6th ser., 2, no. 2 (1980): 205–8. (6)

180. Herdeg, Walter, ed. *Art in the Watermark. Kunst im Wasserzeichen. L'Art du filigrane.* Zurich, 1952. Reprint. 1972. (363)

181. Hermans, C. R. *Papiermerken voorkomende in de rekeningen van het Lieve Vrouwe broederschap te's-Hertogenbosch. Handelingen van het Provincial genootschap van kunsten en wetenschappen in Noord-Brabant.* Bois-le-Duc, 1847. (77)

182. Hermans, J. *Watermerken en papierstructuur.* Amsterdam, 1972.

183. Herring, Richard. *Paper & Paper-Making, Ancient and Modern.* London, 1863. (5)

184. Hind, Arthur M. *Catalogue of Early Italian Engravings Preserved*

in the Department of Prints and Drawings in the British Museum. London, 1910. (72)

185. Hindley, Charles, ed. *The Old Book Collector's Miscellany: Or, a Collection of Readable Reprints of Literary Rareties, Illustrative of the History, Literature, Manners, and Biography of the English Nation During the Sixteenth and Seventeenth Centuries.* London, 1871–1873. (4)

186. Hössle, Fr. von. *Geschichte der alten Papiermühlen im ehemaligen Stift Kempten und in der Reichsstadt Kempten.* Kempten, 1900. (150–64)

187. ———. "Die Uttmer Papiermühle." *Papier-Zeitung* 102 (1906). (8)

188. ———. *Die alten Papiermühlen der Freien Reichsstadt Augsburg sowie alte Papiere und deren Wasserzeichen im Stadt-Archiv und der Kreis- und Stadt-Bibliothek zu Augsburg.* Augsburg, 1907. (316)

189. ———. "Alte Münchener Papiermühle." *Der Papier-Fabrikant* 6 (1908): 1374–75. (2)

190. ———. "Die Papiermühlen im bayerischen Allgäu." *Wochenblatt für Papierfabrikation* 25 (1908): 1985–2006. (116)

191. ———. "Wasserzeichen aus Handschriften und Inkunabeln der K. B. Hof- und Staatsbibliothek: unseres Wasserzeichenforschers Friedrich Keinz, +1901, letzte Arbeit." *Der Papier-Fabrikant* (Festheft) 7 (1909): 61–68. (41)

192. ———. "Wasserzeichen xylographischer Werke der Kgl. Bayr. Hof- und Staatsbibliothek." *Der Papier-Fabrikant* (Festheft) 8 (1910): 32–38. (37)

193. ———. "Wasserzeichen alter Papier des münchener Stadtarchives." *Der Papier-Fabrikant* (Festheft) 9 (1911): 69–75. (17)

194. ———. "Die 4 als Wasserzeichen, Handels- und Hausmarke der alten Papiermacher." *Der Papier-Fabrikant* (Festheft) 10 (1912): 21–26.

195. ———. "Jahreszahlen als Wasserzeichen." *Papier-Zeitung* 38 (1913): 2735–36, 2914–16, 2952. (15)

196. ———. "Karneval im Verborgenen, oder der Harlekin im Papier." *Der Papier-Fabrikant* (Festheft) 12 (1914): 34–39. (22)

197. ———. "Alte Salzburger Papiermühlen." *Zentralblatt für die Papierindustrie* (Vienna) 33 (1915): 309–12. (11)

198. ———. *Württembergische Papiergeschichte: Beschreibung des alten Papiermacher-Handwerks, sowie der alten Papiermühlen im Gebiet des Königreichs Württemberg.* Biberach a.d. Riss, 1925.

(213). Supplement in *Wochenblatt für Papiergeschichte* 41 (1930): 1247.

199. ————. "Alte Papiermühlen-Gründungen im ehemaligen Grossherzogtum Baden." *Zellstoff und Papier* (Berlin) 8 (1928): 405–14. (14)

200. ————. "Alte Papiermühlen der Provinz Schlesien." *Der Papier-Fabrikant* 33 (1935): 9–13, 19–22, 38–40, 221–23, 227–28, 236–38, 246–48, 260–62, 268–71, 279–80, 293–96, 357–59, 369–72, 427–31, 446–48; 34 (1936): 14–15, 29–31, 37–39, 115–19, 163–66, 302–4, 310–12, 333–35, 340–42; 36 (1938): 73–79, 93–95, 101–3. (93)

201. Holman, D. Kern. *Autograph Musical Documents of Hector Berlioz, c. 1818–1840.* (Ph.D. diss., Princeton Unversity, 1974) Ann Arbor, 1979.

202. Horodisch, A. "On the Aesthetics of Watermarks." In *Briquet Album,* edited by Emile J. Labarre. Hilversum, 1952. (72)

203. Hubay, Ilona. *Missalia Hungarica.* Budapest, 1938. (33)

204. Hudson, Frederick. "Wasserzeichen in Händelschen Manuskripten und Drücken (Wasserzeichen in Verbindung mit anderem Beweismaterial zur Datierung der MSS und frühen Drucke G. F. Händels)." In *Handel-Konferenzbericht 1959.* pp. 193–206. Leipzig, 1961.

205. ————. "Concerning the Watermarks in the Manuscripts and Early Prints of G. F. Handel." *Music Review* 20 (1959): 7–27. (19)

206. ————. "The Earliest Paper Made by James Whatman the Elder (1702–1759) and Its Significance in Relation to G. F. Handel and John Walsh." *Music Review* 38 (1977): 15–32. (13)

207. ————. "The New Bedford Manuscript Part-Books of Handel's Setting of 'L'Allegro.' " *Notes: The Quarterly Journal of the Music Library Association* 33 (1977): 531–52. (2)

208. Hunter, Dard. "Ancient Paper-Making." *Miscellany* 2, no. 4, (1915): 67–75. (2)

209. ————. *Handmade Paper and Its Watermarks, a Bibliography.* New York, 1916. Reprint. 1967. Reprinted in article form in *Paper* 20, no. 12 (1917): 20–26.

210. ————. "Symbolism of Paper Markings." *Paper* 33, no. 9 (1923): 3–6. (11)

211. ————. *Papermaking Through Eighteen Centuries.* New York 1930. Reprint. 1971. (188)

212. ————. "Old Watermarks of Animals." *Paper* 28 (1921): 12–15, 25. (22)

213. ———. "The Use and Significance of Ancient Watermarks." *Paper & Printing Digest* (December 1936): 3–7; (January 1937): 3–7; (February 1937): 3–14; (March 1937): 3–5; (April 1937): 3–4. Reprinted from *Papermaking Through Eighteen Centuries*.

214. ———. "Romance of Watermarks." *Printing, Selling and Production* 1 (1938): 9–12, 36–38. (1)

215. ———. "The Watermarking of Portraits Ancient and Modern." *Indian Print and Paper* 5 (1940): 22–25. (8)

216. ———. "Ohio's Pioneer Paper Mills." *Paper Industry* 28, no. 1 (1946): 98, 100, 102, 104.

217. ———. *Papermaking by Hand in America, 1690–1811*. Chillicothe, Ohio, 1950.

218. ———. *Papermaking in Pioneer America*. Philadelphia, 1952. Reprint. 1981 (Expands on the 1950 publication of *Papermaking by Hand in America*) (23)

219. Hunter, Joseph. "Specimens of Marks Used by the Early Manufacturers of Papers, as Exhibited in Documents in the Public Archives of England." *Archaeologia* 37 (1857): 447–54. (30)

220. Hupp, Otto. *Ein Missale speciale, Vorläufer des Psalteriums von 1457*. München-Regensburg, 1898. (3)

I

221. Indra, B. "Opavská papírna." In *Opava: sborník k 10. výročí osvobozeni města*. pp. 165–90. Ostrava, 1956. (33)

222. Irigoin, J. "Les Types de formes utilisés dans l'orient méditerranéen (Syrie, Égypte) du XIe au XIVe siècle." *Papiergeschichte*, 13 (1963): 18–21.

J

223. Jacobs, Eduard. *Geschichtsquellen der Provinz Sachsen*. 1883. (4)

224. Jaffé, Albert. "Die ehemaligen Papiermühlen im heutigen Bezirksamt Pirmasens und ihre Wasserzeichen." *Der Papier-Fabrikant* 26 (1928): 565–70. (7)

225. ———. "Die Neu- oder Apostelmühle bei Rodalben." *Papier-Fabrikant* 26, H. 38 (1928). (3)

226. ———. "Zur Geschichte des Papiers und seiner Wasserzeichen. Eine kulturhistorische Skizze unter besonderer

Berücksichtigung des Gebiets der Rheinpfalz." *Pfälzische Heimatkunde* 26 Jg., H. 3/4 (1930): 71–97. (61)

227. ———. *Die Papierindustrie in den kurpfälzischen Stammlanden unter Kurfürst Carl Theodor.* Pirmasens, 1935. (84)

228. ———. *Geschichte der Papiermühlen im ehemaligen Herzogtum Zweikrüken, mit besonderer Berücksichtigung der allgemeinen Papiergeschichte, sowie der Entwicklungsgeschichte der rheinpfälzischen Papierindustrie und der Wasserzeichentechnik.* Pirmasens, 1933. (136)

229. Jakó, Zs. "Az erdélyi papírmalmok feudálizmuskori történetének vazlata (XVI–XVII. század)" (A Brief Historical Sketch of Transylvanian Paper-Marks During the Feudal Period). *Studia Universitatis Babeş-Bolyai* (Cluj), Ser. historia (1962): 59–81. (17)

230. Janot, Jean Marie. *Les Moulins à papier de la région Vosgienne.* 2 vols. Nancy, 1952. (361)

231. Jansen, Hendrik J. *Essai sur l'origine de la gravure en bois et en taille-douce, et sur la connaissance des estampes des XVe et XVIe siècles.* 2 vols. Paris, 1808. (287)

232. Johnson, Douglas. *Beethoven's Early Sketches in the "Fischhof Miscellany," Berlin Autograph 28.* (Ph.D. diss., University of California, Berkeley, 1977). Ann Arbor, 1980. (See vol. 2, pp. 264–93 for watermarks.)

233. Jones, Horatio Gates. "Historical Sketch of the Rittenhouse Paper-Mill, the First Erected in America, A.D. 1690." *Pennsylvania Magazine of History and Biography* 20, no. 3 (1896): 315–33. (1)

K

234. Kamanin, I., and O. Vitvits'ka. *Vodjani znaky na paperi ukrajins'kyh dokumentiv XVI i XVII vv. (1566–1651).* Kiev, 1923. (1,336)

235. Karlsson, K. K. "Watermarks in Hand-Made Paper in Finland." *Paperi Puu* 44, no. 9 (1962): 459–61, 465–72.

236. Kazmeier, A. W. "Wasserzeichen und Papier der zweiundvierzigzeiligen Bibel." *Gutenberg-Jahrbuch*, 1952, pp. 21–29. (16)

237. Keinz, Friedrich. *Die Wasserzeichen des XIV. Jahrhunderts in Handschriften der K. bayerischen Hof- und Staatsbibliothek.* München, 1896. (368)

238. ———. "Über die älteren Wasserzeichen des Papiers und

ihre Untersuchung." *Zeitschrift für Bücherfreunde* 1 (1897): 240–47. (11)

239. ———. "The Earliest Watermarks." *Literary Collector* 8, no. 6 (1904): 160–68. (11)

240. Kelliher, Hilton. "The Early History of the Malory Manuscript." In *Aspects of Malory,* edited by Toshiyuki Takamiya and Derek Brewer, pp. 143–58. Woodbridge, England, and New Jersey, 1981. (3)

241. Kennedy, D. "The Hibernia Watermark." *Paper Maker* 30: no. 1 (1961): 33–43.

242. Kirchner, Ernst. "Noch einiges über Ulman Stromer, den ersten deutschen Papiermacher zu Nürmberg." *Papier-Zeitung,* (1892): 2035–36. (6)

243. ———. *Die Papierfabrikation in Chemnitz.* Festschrift zum 750 Jahr. Jubiläum der Stadt Chemnitz. Chemnitz, 1893. (17)

244. ———. *Die Papiere des XIV. Jahrhunderts im Stadtarchive zu Frankfurt a. M. und deren Wasserzeichen, technisch untersucht und beschrieben von Ernst Kirchner.* Frankfurt, a. M., 1893. (153)

245. ———. "Die Papierfabrikation der Rheinprovinz." *Wochenblatt für Papierfabrikation* 40 (1909): 1934–54. (11)

246. ———. "Geschichte der deutschen Papierfabrikation: C. F. Walther, Papierfabriken, Flensburg." *Wochenblatt für Papierfabrikation* 38 (1907): 1054–56. (6)

247. ———. "Die Papierfabrikation in Sektion X der Papiermacherberufsgenossenschaft (Brandenburg, Pommern, Ost-, und Westpreussen, Meckenburg)." *Wochenblatt für Papierfabrikation,* Festheft (1911): 2067–97. (19)

248. ———. "Die Papierfabrikation in den Ländern der Sektion III der Papiermacherberufsgenossenschaft, das Grosshertzogtum Baden und das Reichsland Elsass-Lothringen umfassend." *Wochenblatt für Papierfabrikation* 43 (1912): 1979–2030. (14)

249. Kirchner, Joachim. *Lexikon des Buchwesens.* 4 vols. Stuttgart: 1956. See vol. 4, part 2, p. 448. (1)

250. Kirsch, William. "The raison d'être of Mediaeval Papermarks." *Baconiana* 1, ser. 3 (1903): 225–35. (21)

251. Klepikov, S. A. "Bumaga s filigranyu 'gerb goroda Amsterdama.'" *Zapiski Otdela Rukopisei Gos. Biblioteki SSSR im V. I. Lenina* (Moscow) 20 (1958): 315–52. (52)

252. ———. *Filigrani i štempeli na bumage russkogo i inostrannogo proizvodstva XVII-XX veka.* Moscow, 1959.

253. ———. "Bumaga s filigran'yu 'golova shuta (foolscap)': ma-

teriali̇ dlya datirovki rukopisni̇kh i pechatni̇kh dokumen-tov" (Paper With the Foolscap Watermark: Materials for Dating Manuscript and Printed Documents). *Zapiski Otdela Rukopisei Gos. Bibliotekī SSSR im V. I. Lenina* (Moscow) 26 (1963): 405–78. (105)

254. ———. "The Horn Watermark: A Preliminary Classification of Early Varieties, 1314–1600." In *VII International Congress of Paper Historians, Communications*, pp. 63–82. Oxford, 1967. (60)

255. ———. *Filigrani na bumage russkogo proizvodstva XVIII-nachala XX veka* (Watermarks in Russian Papers from the 18th to the Beginning of the 20th Century). Moscow, 1978. (1,378)

256. Klier, Karl M. *Schreibpapier-Wasserzeichen aus dem heutigen burgenländischen Raum von 1703–1710*. Eisenstadt, 1965. (94)

257. Klingelschmidt, Frz. Th. "Eine kurmainzer Papiermühle (I)." *Philobiblon* 9 (1936): 300–302. (1)

258. ———. "Kurmainzer Papiermühle (II)." *Philobiblon* 13 (1940): 96–104. (9)

259. Kohtz, Hans. *Ostpreussische Papierfabrikation*. Stallüpönen, 1935. (51)

260. ———. "Ostpreussische Adler-Wasserzeichen im Wandel der Zeit." *Papiergeschichte* 1 (1951): 47–53. (25)

261. Koning, Jacobus. *Verhandeling over den oorsprong, de uitvinding, verbetering en volmaking der boekdrukkunst*. Haarlem, 1816. (23)

262. ———. *Bijdragen tot de geschiedenis der boekdrukkunst*. Haarlem, 1818–20. (10)

263. Kotte, H. "How the Watermark Arrived in the World." *Allgemeine Papier-Rundschau* 3 (1950): 90–93. (6)

264. ———. "Die Schlopper Papiermühle. Von auf und ab eines Papiermachergeschlects." *Der Papiermacher* 11 (1952): 14–15; 12: 13–14. (7)

265. Kuhne, H. "Conrad Ruhel, Buchhändler und Erbauer einer Wittenberger Papiermühle im 16. Jahrhundert." *Wochenblatt für Papierfabrikation* 92 (1964): 768–73. (3)

266. ———. "Das Schicksal der ersten Wittenberger Papiermühle in der Mitte des 16. Jahrhunderts." *Wochenblatt für Papierfabrikation* 92 (1964): 350–51. (2)

L

267. Labarre, Emile J. *A Dictionary of Paper and Paper-Making Terms, With Equivalents in French, German, Dutch, and Italian.*

Amsterdam, 1937: Rev. ed. London and Amsterdam, 1952. (233) English addenda by Allan Stevenson, *Library,* 5th ser., 9:59–63. Supplement prepared by E. G. Loeber (*Supplement to E. J. Labarre's Dictionary and Encyclopædia of Paper and Paper-Making . . .* , Amsterdam, 1967).

268. ———. *A Short Guide to Books on Watermarks.* Hilversum, 1955. Reprinted in *The Nostitz Papers,* pp. xxxvii–xlii; and in *Philobiblon* 1 (1957): 237–51.

269. ———. "Bücher über Wasserzeichen: eine Bibliographie." *Philobiblon* 1 (1957): 237–51. (20)

270. Labarre, Emile J., gen ed. *The Briquet Album: A Miscellany on Watermarks, Supplementing Dr. Briquet's Les Filigranes, by Various Paper Scholars.* (Monumenta Chartæ Papyraceæ Historiam Illustrantia, II) Hilversum, 1952; Chapel Hill. 1981. (92)

271. ———. *The Nostitz Papers: Notes on Watermarks Found in the German Imperial Archives of the 17th and 18th Centuries, and Essays Showing the Evolution of a Number of Watermarks.* (Monumenta Chartæ Papyraceæ Historiam Illustrantia, V) Hilversum, 1956. (766)

272. Lacombe, H. *Symbolisme du filigrane au fou.* Angouleme, 1958. (5)

273. Lanteff, Ivan. *A vízjegyekröl.* Szent-Pétervár, 1824. (145)

274. La Rue, Jan. "Wasserzeichen." In *Die Musik in Geschichte und Gegenwart; Allgemeine Enxyklopädie der Musik,* vol. 14 (1968), cols. 265–67. (25)

275. La Rue, Jan, and J. S. G. Simmons. "Watermarks." In *The New Grove Dictionary of Music and Musicians,* 20: 228–31. London, 1980. (6)

276. Lassen, T. *Danske og norske historike vandmaerker.* Odense, 1922.

277. Laszlo, Szonyi J. *14 szazadbeli papiros-okleveleink vizjegyei.* Budapest, 1908. (233)

278. Latour, A. "Het watermerk." *De Papierwereld* 3 (1949): 442–43. (2)

279. Laucevičius, Edmundas. *Popierius Lietuvoje XV–XVIII a* (Paper in Lithuania in the 15th–18th Century). 2 vols. Vilnius, 1967. (4,277)

280. Lechi, Luigi. *Della tipografia bresciana nel secolo decimoquinto.* Brescia, 1854. (20)

281. Le Clert, Louis. *Le Papier: Recherches et notes pour servir à l'histoire du papier, principalement à Troyes et aux environs depuis le XIVème siècle.* 2 vols. Paris, 1926. (670)

282. Lehrs, Max. *Geschichte und kritischer Katalog des deutschen,*

niederländischen und französichen Kupferstichs im XV. Jahrhundert. Wien, 1908–34. Reprint (9 vols.) New York, 1969. (560). A single volume containing reproductions of all the illustrations in *Geschichte und kritischer Katalog,* New York, 1969.

283. Lemon, R. "A Collection of Watermarks by the Late Mr. R. Lemon of the Record Office." In Rev. Dr. Scott and Samuel Davey, *A Guide to the Collector of Historical Documents, Literary Manuscripts and Autograph Letters, etc.* London, 1891. (138)

284. Lenz, Hans. *El papel indígena Mexicano: historia y supervivencia.* Mexico, 1948. (1)

285. Lenz, Hans, and Federico Gómez de Orozco. *La industria papelera en Mexico.* Mexico, 1940. (102)

286. Lewis, John. *The Life of Mayster Wyllyam Caxton . . . the First Printer in England.* London, 1737. (20)

287. Likhachev, N. P. *Bumaga i drevnejšie' bumažnye mel'nicy v Moskovskom gosudarstve.* Saint Petersburg, 1891. (783)

288. ———. *Paleografičeskoe značenie bumažnyh vodjanyh znakov.* 3 vols. Saint Petersburg, 1899. (4,258)

289. Lindt, Johann. *The Paper-Mills of Berne and Their Watermarks, 1465–1859.* (Monumenta Chartæ Papyraceæ Historiam Illustrantia, X) Hilversum, 1964. (787)

290. Lorentzen, T. "Die Papiermacherei in der vormaligen Grafschaft Henneberg. I Teil: Schleusingen." *Aus der Praxis des Papiermachers* (Schotten) 4 (1940): 127–56. (110)

M

291. Magee, James F., Jr. "Watermarks of Early American Paper Makers." *Paper Trade Journal,* 24 May 1934, pp. 43–44. (9)

292. ———. "Pennsylvania Colonial Paper Mills and Watermarks." *Proceedings of the Numismatic and Antiquarian Society of Philadelphia* 32 (1935): 217–26. (9)

293. Maleczyńska, K. *Dzieje starego papiernictwa śląskiego* (History of Earlier Papermaking in Silesia). Wrocław-Warszawa-Kraków, 1961. (25)

294. ———. *Dzieje starego papieru (Ksiazkio książce).* Warsaw, 1974.

295. Manzoni, G. "Annali tipografici torinesi del secolo XV." *Miscellanea di Storia Italiana* 4 (1863). (33)

296. Marabini, Edmund. *Die Wasserzeichen in Büttenpapieren des 14–19 Jahrhunderts.* München, 1889. (34)

297. ———. *Bayerische Papiergeschichte*. I Teil: *Die Papiermühlen im Gebiete der weiland Freien Reichsstadt Nürnberg*. Nürnberg, 1894. (85); II Teil: *Die Papiermühlen im ehemaligen Burggrafenthum Nürnberg, den brandenburg-ansbach- und bayreutischen Landen*. München, 1896. (74). Listed elsewhere under the title, *Papiergeschichte der Reichsstadt und des Burggrafenthums Nürnberg*.

298. [Entry deleted.] Cf. #410.

299. Marmol, Baron F. del. *Dictionnaire des filigranes, classés en groupes alphabétique et chronologique*. Namur, 1900. (195)

300. Martin, E. "Watermarks on Paper: Their Value as Criminalistic Evidence." *Internationale Kriminalistische Revue* 16 (1961): 205–11. (5)

301. Martin, Horace R. "Watermarks." *Paper Making and the Printer* 76, no. 2 (1957): 6–16. (6)

302. Matsyuk, O. "Do istoriyi ukrayins'kîkh papiren' XVI st. ta yikh vodyanîkh znakiv" (On the History of 16th-Century Ukrainian Paper-Mills and Their Watermarks). *Naukovo-informatsiīniī byuleten' Archivnoho upravlinnya URSR* (Kiev) 16, no. 5 (1962): 10–21. (15)

303. ———. "Vodyani znakî deyakîkh ukrayins'kîkh papiren' XVI-pochatku XX st" (Watermarks of Some Ukranian Paper-Mills from the 16th to the 20th Century). *Naukovo-informatsiīniī byuleten' Archivnoho upravlinnya URSR* (Kiev) 18 no. 1 (63) (1964): 13–26. (50)

304. ———. "Vodyani znaki na paperi drukiv Ivana Fëdorova" (Watermarks in Books Printed by Ivan Fëderov). *Naukovo-informatsiīniī byuleten' Archivnoho upravlinnya URSR* (Kiev) 18, no. 1 (63) (1964): 13–26. (50)

305. McCorison, Marcus. "Vermont Paper Making." *Vermont History* 31 (1963): 209–45. (8)

306. McElroy, Ken. "Swords and Sabres: Paper and Pens: A Note on Watermarks as Historical Evidence." *Direction Line* 7 (1978): 19–31. (11)

307. Meder, Joseph. *Dürer-Katalog: ein Handbuch über Albrecht Dürers Stiche, Radierungen, Holzschnitte, deren Zustände Ausgaben und Wasserzeichen*. Wien, 1932. Reprint. New York, 1971. (361)

308. Meldau, Robert. "Reichsprivilegien für Wasserzeichen." *Gutenberg-Jahrbuch* 12 (1937): 13–17. (2)

309. ———. "Zur Bedeutung der Hand als Wasserzeichen." *Gutenberg-Jahrbuch* 15 (1940): 41–50. (6)

310. Mena, Ramon. *Filigranas o marcas transparentes en papeles de Neueva España, del siglo XVI.* Mexico, 1926. (21)

311. Mentberger, V. "Two Recently Discovered Paper Trademarks." *Papír a celulosa* 14, no. 6 (1959): 140–41. (2)

312. Mezey, László. *Codices latini medii aevi Bibliothecae Universitatis Budapestinensis quos recensuit Ladislaus Mezey. Accedunt tabulae quae scripturas sub datis exaratas et aliae quae signa chartarum exhibent quas posteriores collegit et notis auxit Agnes Bolgár.* Budapest, 1961. (292)

313. Midoux, Etienne and Auguste Matton. *Étude sur les filigranes des papiers employés en France aux XIVe et XVe siècles.* Paris, 1868. (600)

314. Milano, N. *Della fabbricazione della carta in Amalfi.* Amalfi, 1965. (80)

315. Mitterwieser, Alois. "Frühere Papiermühlen in Altbayern und ihre Wasserzeichen." *Gutenberg-Jahrbuch* 7 (1933): 9–22. (22)

316. ———. "Die alten Papiermühlen bei Landsberg am Lech." *St. Wiborda* (Westheim) 5 (1938): 90–93. (5)

317. ———. "Die alten Papiermühlen van Landshut an der Isar und Braunau am Inn." *Gutenberg-Jahrbuch* 14 (1939): 31–38. (13)

318. ———. "Die alten Papiermühlen Münchens." *Gutenberg-Jahrbuch* 15 (1940): 25–34. (7)

319. Mocarski, *Książka w Toruniu* (Toruń Paper-Mill). Toruń, 1934. (6)

320. Monceaux, Henri. *Les Le Rouge de Chablis: calligraphes et miniaturistes, graveurs et imprimeurs. Étude sur les débuts de l'illustration du livre au XVe siècle.* 2 vols. Paris, 1896. (600) Reprinted from *Bulletin de la Société des Sciences Historiques et Naturelles de l'Yonne.* Vols. 48–50. Auxerre, 1894–96.

321. Mosin, V. "Vodeni znaci najstarijih srpskih štampanih knjiga" (The Watermarks in the Oldest Serbian Printed Books). *Zbornik. Muzej primenjene umetnosti* (Belgrade) 11 (1967): 7–28. (93)

322. ———. *Anchor Watermarks.* Amsterdam, 1973. (Monumenta Chartæ Papyraceæ Historiam Illustrantia, XIII) (2,847)

323. Mosin, V., and Mira Grozdanović-Pajić. "Das Wasserzeichen 'Krone mit Stern und Halbmond.'" *Papiergeschichte* 13 (1963): 44–52. (131)

324. ———. *Agneau Pascal.* Belgrade, 1967. (338)

325. Mosin, V., and S. M. Traljić. *Vodeni znakovi XIII. i XIV. vijeka. Filigranes des XIIIe et XIVe ss.* 2 vols. Zagreb, 1957. (7,271)

326. Murphy, James L. "Watermark Proves Existence of Early Ohio Paper Mill." *Echoes* (August 1978): 4. (1)

N

327. Nadler, A. "Die Papierbezugugsquellen der Freien Reichsstadt Schweinfurt (1550–1750)." *Miscellanea Suinfurtensia Historica, IV, Veröffentlichungen des historischen Vereins und des Stadtarchivs Schweinfurt* (Schweinfurt), Sonderreihe 6 (1964), 93–118. (25)

328. Nardin, Léon. *Jaques Foillet, imprimeur, libraire et papetier (1554–1649).* Paris, 1906. (16)

328A. Needham, Paul. "Johann Gutenberg and the Catholicon Press." *Papers of the Bibliographical Society of America* 76 (1982): 395–451. (27)

329. Nicolaï, A. *Histoire des moulins à papier du sud-ouest de la France, 1300–1800: Périgord, Agenais, Angoumois, Soule, Béarn.* 2 vols. Bordeaux, 1935. (1,804)

330. ———. *Le Symbolisme chrétien dans les filigranes du papier.* Grenoble, 1936. (57)

330A. Nikander, Gabriel and Ingwald Sourander. *Lumppappersbruken i Finland,* Helsingfors, 1955. (42)

331. Nikolaev, V. *Watermarks of the Mediaeval Ottoman Documents in Bulgarian Libraries.* Sofija, 1954. (1,200)

332. Nordstrand, O. "Vandmærker og vandmærkeforskning: Papirhistoriske noter I." *Fund og Forskning i Det Kongelige Biblioteks Samlinger* 17 (1970): 7–20. (4)

O

333. Oberrheiner, C. "Papeteries et papetiers de Cernay." *Revue d'Alsace* 73 (1926): 30–49. (3)

334. Offenberg, A. K. "The Dating of the Kol Bô: Watermarks and Hebrew Bibliography." *Studia Rosenthal* 6 (1972).

335. Ongania, F. *L'Arte della stampa nel rinascimento italiano.* Venice, 1894. (237)

336. ———. *L'Art de l'imprimerie à Venise.* Venice and New York, 1896–97. (237) [Edited and translated from the Italian edition by M. Le Monnier]

P

337. Pangkofer, J. A., and J. R. Schuegraf. *Geschichte der Buchdruckkunst in Regensburg*. Regensburg, 1840. (8)

338. Perrin, André. *Les Caproni, fabricants de papier à la Serraz (Bourget-du-Lac) et à Divonne aux XVIIe et XVIIIe siècles. Leurs marques et filigranes.*Chambéry, 1892. (110–124)

339. Piccard, Gerhard. "Die Wasserzeichenforschung als historische Hilfswissenschaft." *Archivalische Zeitschrift* 52 (1956): 62–115. (150)

340. ———. "Die Datierung des Missale speciale (Constantiense) durch seine Papiermarken." *Archiv für Geschichte des Buchwesens* 2 (1960): 571–84. (54)

341. ———. "Die Kronen-Wasserzeichen." *Veröffentlichungen der Staatlichen Archivverwaltung Baden-Württemberg. Sonderreihe die Wasserzeichenkartei Piccard im Hauptstaatsarchiv Stuttgart.* Findbuch 1. Stuttgart, 1961. (547)

342. ———. "Rechtsrheinische (badische) Papiermühlen und ihre Beziehungen zu Strassburg," *Archiv für Geschichte des Buchwesens* 4 (1962): 997–1102. (35)

343. ———. "Zur Geschichte der Papiererzeugung in der Reichsstadt Memmingen." *Memminger Geschichtsblätter*, 1963 (Memmingen, 1964), pp. 42–61. (60)

344. ———. "Die Ochsenkopf-Wasserzeichen." *Veröffentlichungen der Staatlichen Archivverwaltung Baden-Württemberg. Sonderreihe die Wasserzeichenkartei Piccard im Hauptstaatsarchiv Stuttgart.* Findbuch 2. 3 vols. Stuttgart, 1966. (3,992)

345. ———. "Papiererzeugung und Buchdruck in Basel zum Beginn des 16. Jahrhunderts." *Archiv für Geschichte des Buchwesens* 8 (1966): 322–25. (35)

346. ———. "Das Wappen der Herzoge von Julich-Berg, Cleve, Mark und Ravensberg." *Papiergeschichte* 18 (1967): 1–14. (28)

347. ———. "Die Turm-Wasserzeichen." *Veröffentlichungen der Staatlichen Archivverwaltung Baden-Württemberg. Sonderreihe die Wasserzeichenkartei Piccard im Hauptstaatsarchiv Stuttgart.* Findbuch 3. Stuttgart, 1970.

348. ———. "Wasserzeichen Buchstabe P." *Veröffentlichungen der Staatlichen Archivverwaltung Baden-Württemberg. Sonderreihe die Wasserzeichenkartei Piccard im Hauptstaatsarchiv Stuttgart.* 3 vols. Findbuch 4. Stuttgart, 1977.

349. ———. "Wasserzeichen Waage." *Veröffentlichungen der Staatlichen Archivverwaltung Baden-Württemberg. Sonderreihe*

die Wasserzeichenkartei Piccard im Hauptstaatsarchiv Stuttgart. Findbuch 5. Stuttgart, 1978.

350. ———. "Wasserzeichen Anker." *Veröffentlichungen der Staatlichen Archivverwaltung Baden-Württemberg. Sonderreihe die Wasserzeichenkartei Piccard im Hauptstaatsarchiv Stuttgart.* Findbuch 6. Stuttgart, 1978.

351. ———. "Wasserzeichen Horn." *Veröffentlichungen der Staatlichen Archivverwaltung Baden-Württemberg. Sonderreihe die Wasserzeichenkartei Piccard im Hauptstaatsarchiv Stuttgart.* Findbuch 7. Stuttgart, 1979.

352. ———. "Wasserzeichen Schlüssel." *Veröffentlichungen der Staatlichen Archivverwaltung Baden-Württemberg. Sonderreihe die Wasserzeichenkartei Piccard im Hauptstaatsarchiv Stuttgart.* Findbuch 8. Stuttgart, 1979.

353. ———. "Wasserzeichen Werkzeug und Waffen." *Veröffentlichungen der Staatlichen Archivverwaltung Baden-Württemberg. Sonderreihe die Wasserzeichenkartei Piccard im Hauptstaatsarchiv Stuttgart.* 2 vols. Findbuch 9. Stuttgart, 1980.

354. ———. "Wasserzeichen Fabeltiere. Greif. Drache. Einhorn." *Veröffentlichungen der Staatlichen Archivverwaltung Baden-Württemberg. Sonderreihe die Wazzerzeichenkartei Piccard im Hauptstaatsarchiv Stuttgart.* Findbuch 10. Stuttgart, 1980.

355. ———. "Wasserzeichen Kreuz." *Veröffentlichungen der Staatlichen Archivverwaltung Baden-Württemberg. Sonderreihe die Wasserzeichenkartei Piccard im Hauptstaatsarchiv Stuttgart.* Findbuch 11. Stuttgart, 1981.

356. ———. "Wasserzeichen. Blatt. Blume. Baum." *Veröffentlichungen der Staatlichen Archivverwaltung Baden-Württemberg. Sonderreihe die Wasserzeichenkartei Piccard im Hauptstaatsarchiv Stuttgart.* Findbuch 12. Stuttgart, 1982.

357. ———. "Wasserzeichen Lilie." *Veröffentlichungen der Staatlichen Archivverwaltung Baden-Württemberg. Sonderreihe die Wasserzeichenkartei Piccard im Hauptstaatsarchiv Stuttgart.* Findbuch 13. Stuttgart, 1983.

358. ———. "Wasserzeichen Frucht." *Veröffentlichungen der Staatlichen Archivverwaltung Baden-Württemberg. Sonderreihe die Wasserzeichenkartei Piccard im Hauptstaatsarchiv Stuttgart.* Findbuch 14. Stuttgart, 1983.

359. Piccard, Gerhard, and L. Sporhan-Kempel. *440 Jahre Papiermacherei.* Freiburg im Breisgau, 1952. (4)

360. Piekosiński, Fr. *Średniowieczne znaki wodne . . . wiek XIV.* Kraków, 1893. (795)

361. ———. *Wybór znaków wodnych z XV stulecia.* Estr. Wiadomości numizmatyczno-archeologiczne, Krakow, 1896. (1,119–1,324)
362. Platbarzdis, A. "Das erste Wasserzeichen zum Kennzeichnen von Banknotenpapier." *Papiergeschichte* 5 (1955): 62–66. (10)
363. Popescu, Paulin. "Mărcile de hîrtie filigranate pe manucrisele slavone din mînăstirea punta." *Din Trecutul Bisericii Noastre* 80 (1962): 938–57. (67)
364. Portal, Ch. *Catalogue des incunables et des livres de la première moitié du XVIe siècle; avec divers facsimiles.* Paris, 1892.
365. Pott, Constance. *Francis Bacon and His Secret Society.* Chicago, 1891. (592)
366. Proteaux, A. *Practical Guide for the Manufacture of Paper and Boards.* Philadelphia, 1866.
367. Ptaśnik, J. *Cracovia impressorum XV et XVI saeculorum.* Leopoli, 1922. (Monumenta Poloniae typographica XV et XVI saeculorum I). For papermakers see pp. 110–26. German translation: *Papiergeschichte* 3 (1953): 62–69. (2)
368. Pulignani, D. "Le antiche cartiere di Foligno." *La Bibliofilia* 11 (1909–10): 102–27. (25)
369. Putman, J. L. *Isotopes.* Baltimore and Middlesex, 1960. (1)

Q

370. Quinat, Karl. "Aus alten Papieren." *Der Papier-Fabrikant* 8 (1909): 920–26. (19)

R

371. Rauchheld, Hans. "Die Wasserzeichen der ersten in Oldenburg gedruckten Bücher." *Der Papier-Fabrikant* 24, H. 43 (1926): 663–64. (7)
372. Rauter, A. *Über die Wasserzeichen der ältesten Leinenpapiere in Schlesien.* Breslau, 1866. (173)
373. Renker, Armin. "Das Wasserzeichen: ein entlegenes Feld bibliophiler Betätigung." *Zeitschrift für Bücherfreunde,* n.s., 19 (1927): 61–66. (19)
374. ———. *Das Buch vom Papier.* Berlin, 1929; Leipzig, 1934, 1936; Weisbaden, 1950, 1952.

375. ———. "Kulturgut des Papiermachers im Mittelalter." *Buch-und Werbekunst,* n.s. 7 (1930): 124–28. (1)

376. ———. "Leben und Schicksal des Wasserzeichenforschers Charles-Moïse Briquet." *Philobiblon* 4 (1931): 16–22, 67–69, 103 ff. (16)

377. ———. "Dr. Karl Theodor Weiss in Mönchweiler und seine Papiergeschichtliche   Sammlung."   *Zeitschrift   für Bücherfreunde* 36 (1932): 147–51. (1)

378. ———. "Büttenpapier als Briefpapier." *Papier-Zeitung* 58 (1933): 1549–50. (2)

379. ———. "Papierliebhaber in allen Zeiten und Ständen." *St. Wiborada* 1 (1934): 68–79. (1)

380. ———. "Alte Papiermühlen einst und jetzt." *Buch- und Werbekunst* 12 (1935): 96–98. (14); *Der Altenburger Papier* 10 (1936): 324 ff. (17)

381. ———. "Sinn und Bedeutung der Wasserzeichen." *Der Papier-Fabrikant,* Technischer Teil 35 (1937): 305–9. (4)

382. ———. "Das Wasserzeichen als Kulturspiegel." *Imprimatur* 7 (1937): 176–89. (33)

383. ———. "Art in the Watermark; Kunst im Wasserzeichen; L'Art du filigrane." *Graphis* 8, no. 39 (1952): 52–61, 92–93. (23)

384. ———. "Düren und sein Papiere." *Der Papiermacher* 4 (1953): 5–7. (2)

385. ———. "Filigranophile." *Philobiblon* 1 (1957): 233–35. (3)

386. ———. "Vom Büttenpapier einst und heute." *Allgemeine Papier-Rundschau,* no. 6 (1960): 280–83. (6)

387. Ridolfi, Roberto. *Le filigrane dei paleotipi: saggio metodologico.* Firenze, 1957. (42)

388. ———. *La stampa in Firenze nel secolo XV.* Firenze, 1958. (29)

389. Ringwalt, J. Luther, ed. *American Encyclopædia of Printing.* Philadelphia, 1871; Reprint. 1977. Microfiche: Louisiana and New York, 1981.

390. Rückert, Georg. "Geschichte der Papiermühlen in Zöschlingsweiler und Schretzheim." *Jahrbuch des historischen Vereins Dillingen,* 1909, pp. 10–28. (42)

S

391. Saffray, H. D. "Un nouvel essai de localisation et de datation de l'incunable GW 644." *Gutenberg-Jahrbuch* 39 (1964): 98–102.

392. Sandred, K. I. *A Middle English Version of the Gesta Romanorum Edited From Gloucester Cathedral MS 22*. Uppsala: Acta Univ. Upsaliensis, Studia Anglistica Upsaliensia, 8, 1971. (13)

393. Sardini, Giacomo. *Esame sui principi della francese ed italiana tipografia*. Lucca, 1796–98. (69)

394. Schacht, G. "Die Papiere und Wasserzeichen des XIV. Jahrhunderts im Hauptstaatsarchiv zu Dresden." *Wochenblatt für Papierfabrikation* 28 (1911): 2127–37. (49)

395. Schenk, Erich. "Wasserzeichen in Beethovenbriefen." *Beethoven Jahrbuch* 5 (1961/64): 7–74.

396. Schiffmann, F. J. "Die Wasserzeichen der datierten Münsterer-Drucke als Zeugen für die Aechtheit eines undatierten." *Jahrbuch für schweizerische Geschichte* 7 (1882). (9)

397. Schmid, W. M. "Niederbayer Papiermühlen." *Niederbayer. Monatsschr.*, H. 1–4 (1919): 1–2. (13)

398. Schmidt, Ch. "Memoire sur les filigranes de papiers employés à Strasbourg de 1342 à 1525." *Bulletin de la Société Industrielle de Mulhouse*, November 1877. (40)

399. Schmitt. P. "Le Moulin à papier de Vieux-Thann (1463–1748)." *Annuaire de la Société d'Histoire des Régions de Thann-Guebwiller*, 1961–64 (Mulhouse, 1965), 1–18. (17)

400. Schoonover, David E. "The Pots of Normandy in English Printed Books." *Direction Line* 7 (1978): 24–31. (8)

401. Schulte, Alfred. "Papiermacherei auf Sanct Antony, 1821–1826." *Werksz. der Gutehoffnungshütte*, no. 19 (1928). (2)

402. ———. "Die ältesten Papiermühlen der Rheinlande." *Gutenberg-Jahrbuch* 7 (1932): 44–52. (3)

403. ———. "Entrümpelung und Papiergeschichte. Rettet alte Handpapiere vor der Vernichtung." *Wochenblatt für Papierfabrikation* 65 (1934): 335–36; *Der Altenburger Papierer* 9 (1935): 11. (1)

404. ———. "Schöpfformen und Doppelformen." *Wochenblatt für Papierfabrikation* 65 (1934): 724–27.

405. ———. "Papiermühlen- und Wasserzeichenforschung." *Gutenberg-Jahrbuch* 9 (1934): 9–27. (6)

406. ———. "Die Wasserzeichen von Schoellershammer." *Der Altenburger Papierer* 9 (1935): 427–38. (16)

407. ———. "Zwei Papiermühlen bei St. Goarshausen." *Wochenblatt für Papierfabrikation* 67 (1936): 919 ff. (1)

408. ———. "Die Papiermühle zu Kiedrich im Rheingau." *Wochenblatt für Papierfabrikation* 67 (1936): 521 ff. (1)

409. ———. "Wasserzeichen aus dem Altpapier." *Der Altenburger Papierer* 11 (1937): 970; supp. 12 (1938): 830–32; 13 (1939):

289 ff. (2); *Der Papier-Fabrikant,* 37 (1939): 267; *Der Alten-burger Papierer* 14 (1940): 68.

410. ———. "Bayerische Papiergeschichte. Ein Nachtrag. (Papier-mühlen Fronberg, Pfaffenhofen, Schmidmühlen, Wild-enau/Oberpfalz)." *Der Papier-Fabrikant* 35 (1937): 493–96. (7)

411. ———. "Die Papiermühle zu Boizenburg a.d. Elbe." *Wochenblatt für Papierfabrikation* 68 (1937): 890. (2)

412. ———. "Zwei Wasserzeichen der Papiermühle Bürgel (Thür)." *Der Altenburger Papierer* 11 (1937): 72. (2)

413. ———. "Altpapier und Papiergeschichte." *Der Waldhofer* 2 (1940): 18 ff. (4)

414. ———. "Die Papiermacherei in Bieberach a.d. Riss." *Wochenblatt für Papierfabrikation* 72 (1941): 385–87. (2)

415. ———. "Die Papiermühle Söflingen bei Ulm um 1469." *Jahrb. Buch und Schrift,* n.F., 4 (1941): 95–104. (1)

416. ———. "Rätsel der Wasserzeichenforschung." *Papier-Geschichte* 1, no. 1 (1951): 9–13. (4)

417. Schwanke, Erich. "Beitrag zur Geschichte der Papier-macherei in Deutsch-Böhmen." *Wochenblatt für Papier-fabrikation* H. 45 (1920): 3175–77. (4)

418. Schweizer. "Die Diessener Papiermühle." *Lech-Isar-Land, Monatsschr. z. Pflege d. Heimatgedankens im Huosigau* 11 (1935): 157–59. (1)

419. Schwenke, Paul. *Johannes Gutenbergs zweiundvierzigzeilige Bibel.* Leipzig, 1923. (6)

420. Secher, V. A. *Fortegnelse over 235 prover paa skiv papirsorter brugte af den danske admenistration fra den 15 til det 19 aarhundrede.* København, 1888. (10)

421. Seitz, May A. *The History of the Hoffman Paper Mills in Mary-land.* Towson, Maryland, 1946. (1)

422. Shorter, A. H. *Paper Mills and Paper Makers in England, 1495–1800.* Hilversum: Monumenta chartae Papyraceae Histo-riam Illustrantia, 6, 1957.

423. Sičynśkyj, V. "Papierfabriken in der Ukraine im XVI.–XVIII. Jahrhundert." *Gutenberg-Jahrbuch,* 1941, pp. 23–29. (6)

424. Siegl, Karl. "Die Egerer Papier-Wasserzeichen." *Unser Egerland* 29 (1925): 3–7. (17)

425. Šilhan, J. *Městské papírny v Litovli: die städtischen Papiermühlen in Litovel (Littau).* Olomouc, 1965. (29)

426. Simmons, J. S. G. "The Leningrad Method of Watermark Reproduction." *Book Collector* 10 (1961): 329–30 + pls.

427. Siniarska-Czaplicka, J. *Znaki wodne papierni Mazowsza w Łat-ach, 1750–1850.* Łódź, 1960.

428. ———. "Paper Mills of the Opaczno District." *Przeglad Papierniczy* 20, no. 12 (1964): 408–11.

429. ———. "Traits caractéristiques des filigranes des moulins polonais." *International Congress of Paper Historians, Communications* 7 (1967): 143–58. (20)

430. Skelton, R. A., Thomas E. Marston, and George D. Painter. *The Vinland Map and the Tartar Relation.* New Haven and London, 1965. (2)

431. Smith, Vincent S. "An Introduction to a Subject of Absorbing Interest to Many Buyers of Printing: Watermarks." *British Printer* 66 (1953): 55–59. (12)

432. Šorn, J. "Starejši mlini za papir na Slovenskem" (The Oldest Slovene Papermills). *Zgodovinski časopis* (Ljubljana) 8 (1954): 87–117. (46). Addendum: 9 (1955): 189–92.

433. Sotheby, S. L. *A Collection of Nearly 500 Facsimiles of the Water-Marks Used by the Early Paper Makers During the Latter Part of the Fourteenth and Early Part of the Fifteenth Centuries.* London, 1840. (Reproductions of eighteen of these marks appear in an article in the *World's Paper Trade Review,* 2 November 1900.)

434. ———. *The Typography of the Fifteenth Century: Being Specimens of the Productions of the Early Continental Printers, Exemplified in a Collection of Fac-similes From One Hundred Works, Together with Their Water Marks.* London, 1845. (600)

435. ———. *Principia typographica. The Block-Books, or Xylographic Delineations of Scripture History, Issued in Holland, Flanders and Germany During the Fifteenth Century, Exemplified and Considered in Connexion with the Origin of Printing. To which is Added an Attempt to Elucidate the Character of the Paper-Marks of the Period.* 3 vols. London, 1858. (500)

436. ———. *Specimen of Mr. S. Leigh Sotheby's Principia typographica, an Extensively Illustrated Work, in Three Volumes, Imperial Quarto, on the Block-Books, or Xylographic Delineations of Scripture History Issued in Holland, Flanders, and Germany, During the Fifteenth Century; on Their Connexion with the Origin of Printing, and on the Character of the Water-Marks in the Paper of the Period.* London 1858. Contains the introduction, list of plates, indexes, and some extra plates from the *Principia;* original title page reads: "Specimen-Notice for the Disposal of Mr. S. Leigh Sotheby's Principia typographica [etc]."

437. ———. *Memoranda Relating to the Block-Books Preserved in the Bibliothèque Impériale, Paris.* London, 1859.

438. ——— and Samuel Sotheby, Sr. *A Collection of Specimens in Tracing or Facsimile of Early Printing; of Tracings of Early Water-Marks on Paper, and of Specimens of Papers with Water-Marks, Brought Together, and Illustrated with Copious Manuscript Notes.* 31 vols. British Museum, b and c, and 319b.

439. Spahr, W. "Juan Gabriel Romani and His Descendants: A Chapter of Spanish Papermaking History." *Wochenblatt für Papierfabrikation* 80, no. 11 (1952): 411–15. (2)

440. Spitta, Philipp. *Johann Sebastian Bach.* 2 vols. Leipzig, 1873, 1880. (See vol. 2, pp. 767–846). English edition translated by Clara Bell and J. A. Fuller-Maitland. 3 vols. London, 1899. (14)

441. Sporhan-Krempel, Lore. "Aus der Geschichte der badischen Papiermühlen und ihre Wasserzeichen." *Wochenblatt für Papierfabrikation* 78, no. 23 (1950): 696–98. (5)

442. ———. "Die badische Papiermühlen und ihre Wasserzeichen." *Das Papier* 4, nos. 17/18 (1950): 351–53.

443. ———. *Ochsenkopf und Doppelturm. Die Geschichte der Papiermacherie in Ravensburg.* Stuttgart-Degerloch, 1953. (73)

444. ———. *Papiermühlen und Papiermacher in Lindau und Oberschwaben.* Lindau, 1957. (26)

445. ———. "Die Papierwirtschaft der Nürnberger Kanzlei und die Geschichte der Papiermacherei im Gebiet der Reichsstadt bis zum Beginn des 30-Jährigen Krieges." *Archiv für Geschichte des Buchwesens* 2 (1958–60); cols. 161–69. (3)

446. ———. "Die früheste Geschichte eines gewerblichen Unternehmens in Deutschland: Ulman Stromers Papiermühle in Nürnberg (mit einem Wasserzeichengutachten von Gerhard Piccard)." *Archiv für Geschichte des Buchwesens* 4 (1961–63), cols. 187–212. (8)

447. ———. "Papiermacherei im ehemaligen Hochstift Osnabrück." *Archiv für Geschichte des Buchwesens* 8 (1966–67), cols. 333–90. (2)

448. Springer, Karl. "Ettlinger Wasserzeichen." *Badische Heimat* 15 (1928): 232–39. (15)

449. Stenger, Erich. "Objektive Festellung von Wasserzeichen." *Der Papier-Fabrikant* 27 (1929): 84–87. (1)

450. Stephenson, Richard W. "The Delineation of a Grand Plan." *Quarterly Journal of the Library of Congress* 36 (1979): 207–24. (1)

451. Stevenson, Allan H. "Shakespearian Dated Watermarks." *Studies in Bibliography* 4 (1951–52): 159–64. (2)

452. ———. "Watermarks are Twins." *Studies in Bibliography* 4 (1951–52): 57–91. (17)
453. ———. *Observations on Paper as Evidence.* University of Kansas Annual Public Lectures on Books and Bibliography, 7. Lawrence, 1961. (6)
454. ———. "Paper as Bibliographical Evidence." *Library,* 5th ser., 17 (1962): 197–212. (24)
455. ———. "The Quincentennial of the Netherlandish Blockbooks." *British Museum Quarterly* 31 (1966–67): 83–87. (4)
456. ———. "Tudor Roses from John Tate." *Studies in Bibliography* 20 (1967): 15–34. (6)
457. ———. *The Problem of the Missale speciale.* London, 1967. (50)
458. Stevenson, Allan, H., ed. *Supplementary Material Contributed by a Number of Scholars to C. M. Briquet, Les Filigranes.* Amsterdam, 1976. (20)
459. Stoppelaar, J. H. *Het papier in de Nederlanden gedurende de middeleeuwen, inzonderheid in Zeeland.* Middelburg, 1869. (259)
460. Sullivan, F. "Little Pitchers in the Big Years." *Paper Maker* 20, no. 1 (1951): 10–19.
461. Szónyi, I. L. "XIV. századbeli papiros-obleveleink vízjegyei" (Watermarks of Hungarian 14th Century Paper Documents). *Magyar könyvszemle,* ser., 2, 15 (1907): 1–16, 123–44, 217–40, 300–331. Book form: Budapest, 1908. (223)

T

462. Tacke, Eberhard. "Beiträge zur Geschichte des Papiers in Niedersachsen und angrenzenden Gebieten (III)." *Papiergeschichte* 5. no. 6 (1955): 79–81. (10)
463. ———. "Eine weitere Papiermühle des ausgehenden 16. Jahrhunderts am Westharz bei Gittelde." *Papiergeschichte* 9 (1959): 48–50. (2)
464. Thayer, Ceil Smith. "Why Is a Watermark?" *Modern Lithography* 36, no. 12 (1968): 43, 45, 48–49. (40)
465. ———. "Patriot Papermakers and Their Watermarks." *Modern Lithography* 38 (1970): 26–31. (17)
466. Thiel, Viktor. "Geschichte der Papiererzeugung und des Papierhandels in Steiermark." *Zentralblatt für die Papierindustrie* 44 (1926): 14–19, 38–42, 60–67, 84–90, 109–13, 134–37, 158–60. (14)
467. ———. "Geschichte der Papiererzeugung und des Pa-

pierhandels in Oberösterreich." *Zentralblatt für die Papierindustrie* 46 (1928): 62–66, 83–85, 105–7, 130–32, 150–55, 174–76, 196–99, 218–20, 240–44, 246–66. (27)

468. ———. "Geschichtliche Nachrichten über die Papiererzeugung in Krain, Görz und Fiume." *Zentralblatt für die Papierindustrie* 49 (1931): 305–8, 327–28, 345–48, 381–84. (17)

469. ———. "Zur Geschichte der Papiererzeugung in Tirol, Vorarlberg und in den 'Vorlanden.'" *Wochenblatt für Papierfabrikation* 68 (1937): Sondernummer, 1–19; 69 (1938): 384–85, 468–69, 507–9. (13)

470. ———. "Geschichte der Papiererzeugung in den Herzogtümern Kärnten und Krain, sowie in der Grafschaft Görz." *Der Altenburger Papierer* 11 (1937): 782–87, 858–63, 955–62, 1052–54. (9)

471. ———. *Geschichte der Papiererzeugung im Donauraum.* Biberach a.d. Riss, 1940. (61)

472. ———. *Die Geschichte der Papiermühle in Stattersdorf. Herausgegeben zur Feier des 150jährigen Familienbesitzes der Stattersdorfer Papier-, Holzstoff- und Zellulosefabriken Matthäus Salzers Söhne, 1798–1948.* Wein, 1948. (34)

473. Timperley, C. H. *A Dictionary of Printers and Printing, With the Progress of Literature, Ancient and Modern.* London, 1839. (5)

474. Todericiu, D. "Filigranele hîrtiilor fabricate în perioada 1539–1841" (Watermarks in Paper Manufactured 1539–1841). *Celuloză si hîrtie* (Bucharest) 11 (1962): 302–5. (40)

475. Traljić, S. M. "Prve kontramarke u talijanskom papiru prema materijalu iz naših arhiva" (The Earliest Countermarks in Italian Papers Preserved in Our Archives). *Zbornik Historijskog Instituta Jugoslavenske Akademije* (Zagreb) 2 (1959): 151–65. (48)

476. ———. "Vodeni znakovi u dokumentima i rukopisima samostana Sv. Marije u Zadru." *Radovi Instituta Jugoslavenske Akademije znanosti i umjetnosti u Zadru* 13–14 (1968): 267–74.

477. Tromonin, K. Ya. *Tromonin's Watermark Album: A Facsimile of the Moscow 1844 Edition,* with additional materials by S. A. Klepikov, edited by J. S. G. Simmons. Hilversum, 1965. (1,827)

478. Tronnier, Ad. "Die Missaldruck Peter Schöffers und seines Söhnes Johann." *Veröffentlichungen der Gutenberg-Gesellschaft* 5–7 (1908): 80–83. Additional material: "Das Papier," pp. 203–5. (27)

479. Tschudin, W. F. *Alter Basler Papiermarken*. Basle, 1954–55. (93)
480. ———. *The Ancient Paper-Mills of Basle and Their Marks*. Hilversum, 1958. (Monumenta Chartæ Papyraceæ Historiam Illustrantia, VII). (429)
481. ———. "The Old Papermills of Switzerland and Their History." *Papermaker* 31 (1962): 39–45; 32, no. 1 (1963): 4–9; no. 2:5–9, 25–33. (5)
482. ———. *Stand der Forschung über die schweizerischen Papiermühlen, Papier- und Kartonfabriken und deren Marken zur Zeit der Schweizerischen Landesausstellung 1964*. Basel, 1964. (7)
483. Tyson, Alan. "New Light on Mozart's 'Prussian Quartets.'" *Musical Times* 116 (1975): 126–30. (6)
484. ———. "'La clemenza si Tito' and Its Chronology." *Musical Times* 116 (1975): 221–27. (6)
485. ———. "Mozart's 'Haydn' Quartets: The Contribution of Paper Studies." In *The String Quartets of Haydn, Mozart, and Beethoven: Studies of the Autograph Manuscripts*, edited by Christophen Wolff, pp. 179–90. Cambridge, Massachusetts, 1980.

U

486. Učastkina, Z. V. *A History of Russian Hand Paper-Mills and Their Watermarks*. Hilversum, 1962. (Monumenta Chartæ Papyraceæ Historiam Illustrantia, IX).
489. Unverricht, Hubert. "Zur Datierung der Bläsersonaten von Johann D. Zelenka." *Die Musikforschung* 15 (1962): 265–68. (2)
488. Urbani, D. *Segni di cartiere antiche*. Venezia, 1870. (150–56)

V

489. Vallet de Viriville. "Notes pour servir à l'histoire du papier." *La Gazette des Beaux Arts* 2 (1859): 222–36. (28)
490. Valls i Subirà, O. *Paper and Watermarks in Catalonia*. 2 vols. Amsterdam, 1970.
491. Vock, Walther E. "Ulman Stromer (1329–1407) und sein Buch. Nachtrag z. hegelschen ausg." *Mitteilungen der Vereinigung für Geschichte der Stadt Nürnberg* 29 (1928): 85–168. (6)

492. Volpicella, Luigi. *Primo contributo alla conoscenza delle filigrane nelle carte antiche di Lucca.* Lucca, 1911 (333)
493. Voorn, Henk. "Het fabeldier als watermerk: de basilisk en de draak." *De Papierwereld* 5 (1950): 99–102. (7)
494. ———. "Het fabeldier als watermerk: de eenhoorn." *De Papierwereld* 5 (1950): 59–62. (3)
495. ———. "Natural History in Watermarks." *Pulp and Paper Magazine of Canada* 56. no. 7 (1955): 93.
496. ———. "Fabulous Beasts in Watermarks." *Paper Maker* 26, no. 2 (1957): 1–9. (11)
497. ———. "Fabulous Beasts in Watermarks: The Basilisk." *Paper Maker* 26 (1957): 19–24. (12)
498. ———. *The Paper Mills of Denmark & Norway and Their Watermarks.* Hilversum, 1959. (Supplement to Monumenta Chartæ Papyraceæ Historiam Illustrantia).
499. ———. *De papiermolens in de provincie Noord-Holland [door] H. Voorn* (De Geschiedenis der Nederlandse Papierindustrie, I). Haarlem, 1960.
500. ———. *De papiermolens in de provincie Zuid-Holland alsmede in Zeeland, Utrecht, Noord-Brabant, Groningen, Friesland, Drenthe, [door] H. Voorn* (De Geschiedenis der Nederlandse Papierindustrie, II). Wormerveer, 1973.

W

501. *Watermarks.* A Collection of Single Leaves, Mostly Blank, Taken from Books Chiefly of the 16th Century, and All Bearing Watermarks. (British Museum)
502. *Watermarks. A Series of Reproductions of Watermarks with Explanatory Notes by Count de Witte.* Brussels, 1912.
503. Weerth, O. "Das Papier und die Papiermühlen im Fürstentum Lippe." *Mitteilungen aus der lippischen Geschichts- und Landeskunde* (Detmold) 2 (1904): 1–130. (107)
504. Weigel, Theodor O. *Collectio Weigeliana: die Anfänge der Druckerkunst in Bild und Schrift: an deren frühesten Erzeugnissen in der Weigel'schen Sammlung erläutert von T. O. Weigel und Dr. Ad. Westermann.* 2 vols. Leipzig, 1866. (100)
505. Weiss, Karl Th. "Die Papiermühle zu Stockach, ihre Geschichte und ihre Wasserzeichen." *Schriften des Vereins für Geschichte des Bodeness* (Lindau) 44 (1915): 14–31. (23)
506. ———. "Deutsche Wappenwasserzeichen." *Der Deutsche Herold* 46 (1915): 99–103, 113–14, 133–37. (62)

507. ———. "Die Bestimmung einer Stuttgarter Handschrift mit Hilfe des verwendeten Papieres und seiner Wasserzeichen." *Gutenberg-Jahrbuch* 28 (1953): 16–24. (5)

508. Weiss, Karl Th. *Handbuch der Wasserzeichenkunde*. Bearbeitet und herausgegeben von Dr. Wisso Weiss. Leipzig, 1962.

509. Weiss, Wisso. "Das Posthorn: ein Beitrag zur Wasserzeichenkunde." *Gutenberg-Jahrbuch* 19–24 (1944–49): 39–46. (11)

510. ———. "Die Wasserzeichenpapiere des Papierwerks Blankenburg im Wandel der Jahrhunderte." *Wochenblatt für Papierfabrikation* 79 (1951): 771–77. (5)

511. ———. *Thüringer Papiermühlen und ihre Wasserzeichen*. Weimar, 1953. (46)

512. ———. "Vom Stempelpapier und seinem Wasserzeichen." *Gutenberg-Jahrbuch* 32 (1957): 26–32. (2)

513. ———. "Eckzier-Wasserzeichen." *Gutenberg-Jahrbuch* 33 (1958): 37–43. (2)

514. ———. "Das Wasserzeichen im alten handgeschöpften Velinpapier." *Gutenberg-Jahrbuch* 36 (1961): 11–17. (3)

515. Wheelright, W. B. "Watermarks." *Paper Maker* 17, no. 2 (1948): 1–6.

516. Wibiral, Fr. *L'Iconographie d'Antoine van Dyck, d'après les recherches de H. Weber*. Leipzig, 1877. (92)

517. Wiener, Lucien. *Étude sur les filigranes des papiers lorrains*. Nancy, 1893. (211)

518. Willcox, Joseph. *Ivy Mills, 1729–1866*. Baltimore, 1911. Philadelphia, 1917.

519. Williamson, K. "Paper Making in Man; the Story of a Lost Industry." *World's Paper Trade Review* 121, no. 23 (1944): 1307–8, 1310, 1342, 1344; no. 24 (1944): 1364, 1367–68, 1404, 1406.

520. Wolf, Eugene K., and Jean K. Wolf. "A Newly Identified Complex of Manuscripts from Mannheim." *Journal of the American Musicological Society* 27 (1974): 371–437. (15)

521. Woodward, David. "Watermark Radiography at the Newberry Library." *Mapline* 15 (1979): 1–2. (1)

522. Wurtembergische, H. "Wasserzeichen." *Papier-Zeitung* 100 (1906). (7)

Z

523. Zonghi, Aurelio. *Le marche principali delle carte Fabrianesi dal 1293 al 1599*. Published with English translation in *Zonghi's*

Watermarks. Fabriano, 1881. (336)

524. ———. *Le antiche carte fabrianese alla Esposizione generale italiana di Torino*. Watermarks reproduced in *Zonghi's Watermarks*. Fano, 1884.

525. Zonghi, Aurelio, and Augusto Zonghi. *Zonghi's Watermarks*. Hilversum, 1953. (Monumenta Chartæ Papyraceæ Historiam Illustrantia, III) (1,887)

526. Zuman, Franz. "Vodní značka průvodním prostředkem." *Památky archeologické* 32 (1921): 260–61. (1)

527. ———. "České filigrány XVI. století." *Památky archeologické* 33 (1923): 277–86. (26)

528. ———. *Papírny Starého města pražského* (The Prague Old Town Paper-Mills). V. Praze, 1927. (15)

529. ———. "České filigrány XVII. století (I)." *Památky archeologické* 35 (1929): 136–66, 442–63. (132)

530. ———. "Papírna trutnovská" (The Trutnov Paper-Mill). *Věstník král. české společnosti nauk, tř. filos. -hist. -jazykopytná* (Prague), 1931, pp. 1–26. (15)

531. ———. *České filigrány XVIII. století*. V. Praze, 1932. (204)

532. ———. *České filigrány z první polovice XIX. století*. V. Praze, 1934. (200)

533. ———. "Die Papiermühle in Schirgiswalde. Ein Beitrag zur Geschichte derselben aus den Prager Archivem." *Der Altenburger Papierer* 10 (1936): 750–55. (2)

534. Zurowski, S. "Znaki wodne papiernictwa wielkopolskiego XVI–XIX w" (Watermarks of Paper-Mills in Great Poland from the 16th-19th Century). *Zeszyty naukowe Uniwersyteta im. Adama Mickiewicza* (Poznań) 57 (1965): 255–316. (40)

6

Techniques of Reproducing Watermarks: A Practical Introduction

David Schoonover

Historians and bibliographers attempting to study watermarks and chain and wire lines as evidence of date and place of paper-manufacture face immediate problems. Researchers readily discern papermarks when they hold a sheet against light, but usually they are unable to record and reproduce their evidence precisely. This article will explain several methods by which researchers may obtain exact copies of papermarks.

Since the early nineteenth century, scholars have relied on visual examination and hand tracings to locate and record papermarks, but these methods are inherently imprecise. Tromonin, Heawood, Churchill, and Briquet spent lifetimes dedicated to amassing thousands of tracings from hundreds of archives, yet their compilations are useful primarily as general guides or illustrated histories. By the mid-twentieth century, Allan Stevenson and other researchers recognized that they needed objective, reliable, precise methods for recording and reproducing papermarks. Stevenson knew that "one weakness of watermarks as evidence is their ambiguity" and he warned that watermarks "lend themselves to subtleties and complications."[1] Filigranists wished to reduce ambiguity in their evidence and monitor such subtleties and complications as they discovered over time. By identifying twin, identical, and variant marks from paper stocks used in editions of books at specific times, then determining a chronology of watermarks through copy after copy, the bibliographers could obtain a basis for relating twin or identical marks from other books or manuscripts. Early in his research, Stevenson recognized the need for comparing paper stocks used by different printers. He announced a major obstacle, however, when he noted that the necessity for exact comparison of papers required carrying record copies of particular marks as a basis for reference. In 1948 no simple technique allowed filigranists to obtain exact copies of papermarks for comparative analysis.

The first advance in the methodology of reproducing watermarks came in the late 1950s. D. P. Erastov, working at the Document Conservation and Restoration Laboratory of the Academy of Sciences at Leningrad, developed a beta-radiographic technique that, as J. S. G. Simmons noted, had "the edge on all existing methods in that it can produce a 'photographic' image of a watermark while entirely eliminating any associated written or printed text."[2] Erastov's method, reported to Western readers by Simmons in 1961 and described later in this article, involves the use of low-energy isotopes to measure differences in paper thickness. Chain lines, wire lines, and watermarks are formed by the wires of the paper mould, which spread paper fibers apart and thus create thinner areas in the sheet. Beta-isotopes can record these variations of paper density on x-ray film with absolute accuracy and clarity. This technique is safe for the researcher and for the materials being examined, the procedures are simple and easily learned, and the process is relatively inexpensive.

When Stevenson published "Paper as Bibliographical Evidence" in 1962, he had just learned about beta-radiography and could recognize its potential: "Unless we can be *sure* of the identity of two marks, beyond any peradventure, we have little chance of solving, convincingly, through paper evidence, such a problem as the date of an incunable."[3] Since direct overlaying of beta-radiographic negatives not only establishes identity of paper moulds and watermarks, but also shows even slight changes in the watermarks, Stevenson knew that this was the method he required. He then proposed the necessary investigation that would link method and materials: "What is needed is a group of bibliographical students who, having thrown off outworn assumptions concerning the value of paper as evidence, will proceed patiently and empirically to examine extant paper moulds and the 'identical' sheets of paper made on such moulds, following the spoor through a number of manuscripts and printed books."[4]

Stevenson was convinced that beta-radiography was bound "to make a large difference in the bibliographical study of watermarks. It may even prove to be the incentive and means needed to make paper investigation available to all."[5] Yet though beta-radiography and other techniques for reproducing papermarks have been available for some years at research libraries in the United States and Europe, these methods have in large measure been neglected, chiefly because scholars are unfamiliar with modern methods of watermark reproduction. I shall now describe three such methods.

Technical Advances

Extensive use of beta-radiography had initially been restricted by the length of time required for each exposure to be made. When the technique was being developed, exposure times ranged from three to twenty hours, depending upon the types of isotope and x-ray film being used. In 1974–75, Warner Barnes of the University of Texas at Austin conducted experiments using two Carbon 14 isotopes at strengths of 3.5 millicuries (mCi) and 40 mCi with two x-ray films, types AA Industrial and Medical No-Screen. This chart shows the greatly improved exposure times:

mCi	Film	Time
3.5	AA Industrial	4 hours
3.5	Medical No-Screen	1 hour
40	AA Industrial	1 hour
40	Medical No-Screen	12 minutes

Thus Barnes's film experimentation reduced radiographical exposure times to a practical level.[6]

Other bibliographers have advocated photographic (rather than radiographic) reproduction of watermarks. In 1976 Robin Alston of the Janus Press, Ilkley, Yorkshire, England, published results of his experiments with high-speed duplicating film to record papermarks. This "Ilkley" technique, essentially a contact printing method, requires only five-second exposures to produce exact-size films of the papermarks and any written or printed material on the sheet of paper being studied.[7]

Thomas L. Gravell has developed a method that uses DuPont's Dylux 503-1 paper, exposed to either ultraviolet or flourescent light, to produce a contact print similar to the Ilkley film, but not requiring darkroom conditions.[8] The Ilkley and Dylux techniques constitute valuable alternatives to beta-radiography.

Beta-Radiography

Advantages and Limitations

Beta-radiography provides a consistently clear image of a paper mould, reproducing the watermark, available sewing dots, and chain and wire lines. The greatest advantage of radiography lies in its capacity to eliminate interference from most written or printed material on a sheet of paper. Thus, with this method researchers may record papermarks that have been obscured heretofore behind lines of ink. The exposed and developed radiograph may be compared easily and quickly to other watermark images, and the original film may be precisely duplicated for exchange and publication by means of contact printing on photographic paper or Kodak LPD4 film.

While many photographic processes require experience with complex equipment, materials, and procedures, beta-radiography may be learned quickly and easily. Researchers using either beta-radiography or the Ilkley technique must have access to darkroom facilities, but most research institutions have photographic services that can accommodate either technique.

Some researchers may consider exposure times of ten to fifteen minutes to be a limitation, but this relatively slow exposure time allows latitude for the radiographer, since differences of seconds or even minutes will still produce a usable image. Finally, because the Carbon 14 sheet emits radiation in all directions, two exposures may be made at once by the method described in "Procedures" below.

Although initial expenditures of approximately $2250 for the Carbon 14 isotope may seem great, there will be no replacement cost since Carbon 14 possesses a half-life of 5,760 years. Other isotopes with shorter half-lives are available at lower costs, but Carbon 14 produced the best results. Neighboring institutions wishing to undertake radiographical research may wish to share the cost of a beta-isotope.

Conditions for receiving, using, and processing radioactive materials are specified by the United States Nuclear Regulatory Commission (NRC), which issues licenses for these materials. Paper researchers in a university or college should enquire whether their institution holds a Broad License from the NRC and any other necessary licenses. The institution's safety or research office can provide this information. Researchers at institutions that do not conduct other radiological projects will need to apply to the NRC, state, and local agencies for licenses.

Isotopes for paper research are in the form of plastic sheets uniformly impregnated with Carbon 14. A 40-millicurie source emits very low-level beta-radiation at the rate of approximately 50 millirems per hour. In contrast, a dental X ray delivers 20 millirems in one-tenth of a second. These low-level beta-emissions involve no radiation hazard to the researcher or to the materials being studied, nor is any radiation transferred from the isotope to the researcher or materials.

Beta-radiography offers a safe, simple, and reliable technique for obtaining precise paper evidence. Allan Stevenson and G. Thomas Tanselle have predicted that the future of watermark studies lies in radiography, and Tanselle asserts that "the student of paper will need to carry with him a Carbon 14 source."[9] It is certainly to be hoped that students of paper will soon find that research institutions can and will answer their requests for radiographic evidence.

Materials

Isotope of C 14 having 40 mCi (millicuries) of activity
Contact: Amersham Corporation, 2636 S. Clearbrook Drive, Arlington Heights, Illinois 60005
Cost: Approximately $2250

Kodak Direct Exposure Film, DEF-5
Contact: Any x-ray supply firm
Cost: Varies according to amount and size of sheets. Price of silver also affects film prices.

Kodak GBX Developer to make 1 gallon
Cost: Approximately $4

Kodak Liquid X-ray Fixer to make 1 gallon
Cost: Approximately $4

Orbit Bath
Contact: Any photographic supply firm

Kodak Photoflo Solution
Contact: Any photographic supply firm

Procedures

Prepare chemicals.
Mix developer as instructed.
Mix fixer: 950 ml H2O + 473 ml "A" + 52 ml "B".
Fill 1 tray with stop-bath solution.
Mix 1 ounce Orbit Bath solution per quart of H2O in tray.
Mix ½ cap of Photoflo with H2O in tray.
Make the radiographic exposure in darkness with safelights.
Place the isotope below the leaf and the film on top of the leaf.
Place sheeets of ¼ inch bevel-edged plate glass above the film and below the isotope to provide firm contact; place a weight on top of the glass.
Double-exposure: If the researcher wishes to record watermarks in successive leaves of a book or in a pair of loose sheets of paper, dual exposures may be made sumultaneously, since the isotope sheet radiates evenly from each side. The materials should be arranged in this order: weight, glass, film, paper, isotope, paper, film, and glass. Exposure time will be the same as required for a single exposure.

Allow approximately 10–12 minutes for the exposure, depending upon thickness of paper
Develop film.

Place film in developer for 2½ minutes, agitating.
Rinse film in stop bath for 30 seconds.
Place film in fixer for 6 minutes, agitating.
Immerse film in Orbit Bath for 5 minutes.
Wash film in circulating water bath for 30 minutes.
Immerse film in Photoflo for 30 seconds.
Hang film to dry.
Complete Radiographical Data Sheet.
Label the exposed and developed film.
Compare successful results with existing films or albums.

Ilkley Technique

Advantages and Limitations

The Ilkley technique may be performed simply and quickly. The film may be superimposed with other images over a light table for comparative study. The original film may also be quickly and accurately duplicated for exchange and publication by contact printing on photographic paper or LPD4 film.

Because the process records any obstruction on the surface of the paper being studied, the watermark may be obscured. However, researchers may wish to record the text as well as paper structure. The Ilkley technique is particularly useful on papers that have relatively little surface obstruction, such as music, maps, or lightly written manuscripts.

Materials

The Ilkley Technique requires simple materials and equipment that are readily available:

KODAK Precision Line Film LPD4, in 8″ × 10″ sheets
Kodalith Developer
Kodak Rapid Fixer
Kodak or other stop bath
Orbit Bath solution
Kodak Photoflo Solution
Five processing trays, 11″ × 14″
One circulating wash tub
Two sheets bevel-edged plate glass, 11″ × 14″ × ¼″
Extendable desk lamp with 25-, 40-, or 60-watt bulb
Besseler or other model photographic enlarger, 105 mm lens
Photographic timer to be connected to lamp or enlarger

Procedure

Locate the watermark by visual examination with the aid of a light table. Raise the enlarger or desk lamp to maximum height (approximately 27 inches), set the timer connected to the enlarger at 5 seconds. Under safelight conditions, remove one sheet of LPD4 film, cut the film to desired size (approximately 4″ × 5″). Taking care not to strain the binding or paper of the work being studied, place one piece of bevel-edged plate glass under the leaf, place the film with emulsion side up beneath the watermark, then cover the leaf with a second piece of glass to provide firm contact between film and paper. The "sandwich" thus formed will be in this order: glass, paper, film, glass. Activate the timer to expose the paper and film to light for 5 seconds. When the exposure is complete, remove the film and process it manually in the following steps:

Place film in Kodalith Developer at 68 degrees F., agitating regularly, for one and one-quarter minutes.
Rinse film in stop bath for 30 seconds.
Place film in Kodak Rapid Fixer at 68 degrees F., agitating regularly, for 2 minutes.
Immerse film in Orbit Bath solution for 5 minutes.
Wash film in circulating water bath for 10 minutes.
Immerse film in Photoflo Solution for 30 seconds to prevent streaking.
Hang film to dry.

After film has dried, label the film, complete a watermark data sheet, and compare the successful result with existing watermark albums. This film is an exact-size positive reproduction of the paper, recording the watermark with sewing dots, chain and wire lines, any variations in paper structure such as paper repairs or thinning of the surface, and any written, printed, or engraved material on either side of the leaf.

Duplicate Negatives, Contact Prints, and Enlargements

Materials

Developed beta-radiograph or Ilkley film
Kodak Precision Line Film LPD4
High-contrast photographic paper
Photographic chemicals (developer, stop bath, fixer, Photoflo)

Procedure

Films produced by beta-radiography or the Ilkley technique may be duplicated quickly and easily, either on film or paper. To make a duplicate film, place the original film on top of a piece of LPD4 film, then place both films in a contact printer or between sheets of clean plate glass. If an enlarger is to be used as the light source, set the aperture to f11 and allow a 10-second exposure. Density of the original film will necessitate adjusting the enlarger's aperture or timer to produce the best duplicate. Test strips should be exposed at different apertures or times before a full-size piece of film is used. After the proper exposure has been determined and made, the film is processed in developer, stop bath, and fixer, then rinsed in a circulating tub of water.
Original films also produce excellent contact prints and enlarge-

ments on photographic paper. To make a contact print, place the original film on top of a piece of paper, then place the pieces in a contact printer or between sheets of clean plate glass. If an enlarger is used as the light source, set the aperture at f11 and allow a 10-second exposure. Again, density of the film will necessitate adjusting the aperture or timer to produce the best print; test strips should be exposed at different apertures or times before full-size paper is used. When the exposure is completed, the paper is processed in developer, stop bath, and fixer, then rinsed in a circulating tub of water.

Enlargements may be useful for illustrations, lectures, or detailed examination. Choose a negative carrier of the proper size, adjust the enlarger for aperture, focus, and time of exposure. Varied contrast in the enlargement can be obtained by adjusting the aperture. When the proper exposure has been determined and completed, the film or paper may be processed as described above.

Dylux Process

Advantages and Limitations

The great advantage of the Dylux process lies in its being performed under lighted conditions, while radiography and the Ilkley technique require photographic darkroom conditions. Dylux processing time, ranging from one to eight minutes, is faster than most radiography, yet slower than the Ilkley technique.

The resulting Dylux image, on paper, must be reproduced carefully for inclusion in watermark albums. Photographic distortion may occur, since the Dylux paper itself cannot be used to make a contact print or negative. This consideration is important, since watermark images should, if possible, be reproduced in original size if they are to be compared or reproduced in albums.

Materials

DuPont Dylux 503-1 Paper
Fluorescent lamp equipped with either Super Diazo or blue daylight tubes
Ultraviolet lamp
Two sheets of bevel-edged plate glass

Procedure

Locate the watermark by visual examination with the aid of a light table. Under ordinary electric room light, but away from sunlight, remove the Dylux paper from its container. Place the Dylux sheet under the watermark, with the yellow-coated side in contact with the paper. To obtain close contact between the two sheets of paper, place the materials between two pieces of plate glass. The exposure is made with the blue daylight or Super Diazo fluorescent lamp, and times will vary according to (1) type of lamp used, (2) thickness of the paper being studied, and (3) amount of writing or printing on the surface of the paper. Mr. Gravell has reported that the Super Diazo lamp requires exposures from 1 to 5 minutes, while the less expensive blue daylight lamp requires from 3 to 8 minutes. With the fluorescent lamp approximately 2 inches from the paper, begin the exposure. When the exposure is complete, remove the Dylux sheet. Next, develop the latent image by exposing the Dylux sheet to ultraviolet light, using a lamp about 12 inches above the Dylux sheet. As the print develops, the yellow coating will turn blue, with chain and wire lines and watermark appearing in contrasting white. When the image has been fully developed, the Dylux sheet should be fixed by exposing it to visible light.

Conclusion

Techniques are now available that can provide evidence for areas of research that Allan Stevenson suggested years ago: histories of paper mills and the paper trade; studies in paper economics to make understandable the part paper has played in the prices of books and of education; and in bibliographical problems using paper evidence in conjunction with typographical evidence. As more researchers in filigranology request and require the results of paper analysis, research institutions should begin to take advantage of the methods described in this article. It is to be hoped that the article will encourage scholars to undertake such studies.

Notes

Northwestern University Library's Scholar-Librarian Program, funded by The Council on Library Resources and The National Endowment for the Humanities, provided primary research support for this project; additional assistance came from Northwestern University Library and the Beinecke Rare Book and Manuscript Library, Yale University. Watermarks reproduced by beta-radiography and Ilkley techniques are taken from printed books in the Special Collections Department, Northwestern University Library, Evanston, Illinois. Watermarks reproduced by the Dylux technique are provided by the courtesy of Thomas L. Gravell. I am particularly grateful for support from Paul Saenger, Russell Maylone, Charles Osburn, John McGowan, Warner Barnes, and Ralph Franklin.

1. Allan Stevenson, "New Uses of Watermarks as Bibliographical Evidence," *Studies in Bibliography* 1 (1948): 179, 182.

2. J. S. G. Simmons, "The Leningrad Method of Watermark Reproduction," *Book Collector* 10 (1961): 329.

3. Allan Stevenson, "Paper as Bibliographical Evidence," *Library* 5th ser., 17(1962): 199.

4. Ibid., p. 211.

5. Allan Stevenson, *The Problem of the Missale speciale* (London: The Bibliographical Society, 1967), 67.

6. Warner Barnes, "Film Experimentation in Beta-Radiography," *Direction Line* 1 (1975): 3–4.

7. Robin Alston, "Reproducing Watermarks," *Direction Line* 2 (1976): 1–3.

8. Thomas L. Gravell, "A New Method of Reproducing Watermarks for Study," *Restaurator* 2 (1975): 94–104.

9. G. Thomas Tanselle, "The Bibliographical Description of Paper," *Studies in Bibliography* 24 (1971): 50.

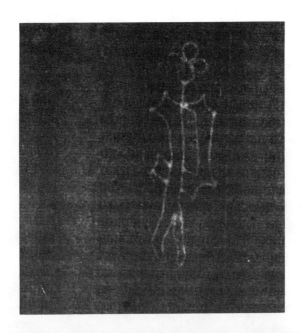

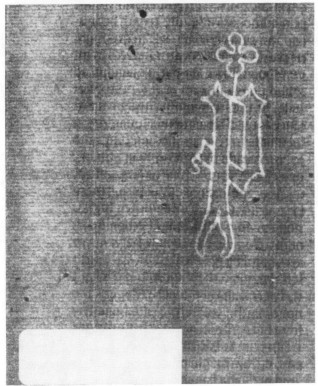

PLATE 1. Jacobus de Voragine, *De Sanctis* (Strassburg: Johann Gruninger, 1484).
Leaves o 7 and z 2, showing twin watermarks, not recorded in Briquet.

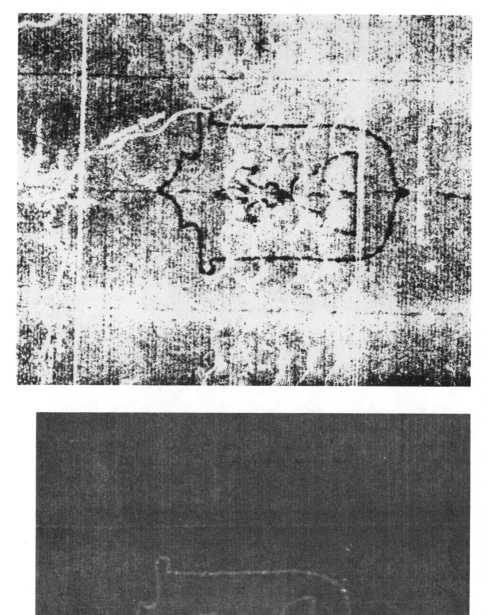

PLATE 2. M. Livio Sanuto, *Geografia* (Venice: Damiano Zenaro, 1588). Map: Africae Tabula VI, showing watermarks reproduced (*left*) by beta-radiography

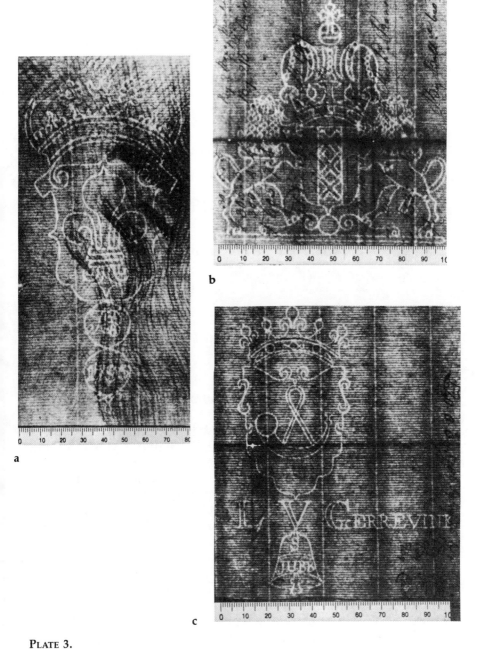

Plate 3.
(a) Arms of Amsterdam: account sheet for Wm. and M. Jones with W. and S. Bernon of Newport, Rhode Island, dated 1750 (Winterthur 72X115).
(b) TB/1795/Shield with fleur: from a page of drawings done by a student of the drawing school at Nazareth, Pennsylvania, dated 1801.
(c) Jubb/S/in a bell: account sheet in "Record Book B" of Old Swedes Church, Wilmington, Delaware, dated 1754.

7

Beethoven's Leonore Sketchbook (Mendelssohn 15): Problems of Reconstruction and of Chronology

Alan Tyson

Wir schwanken zwischen Wahrscheinlichkeit und Vermuthung
—G. Nottebohm (1880)

Introduction

It is unfortunately the case that, with the single exception of the Kessler Sketchbook, now in the Gesellschaft der Musikfreunde in Vienna, there is no sketchbook of Beethoven's that we can be certain is still in the same condition as it was in when it was used by him. Most of them, in fact, have clearly lost some of their leaves. Many sketchbooks show obvious signs of this damage; others, though not offering any direct clues, are still probably far from intact.

It will be plain that any disturbance of the sketch sequence through the loss of leaves limits the value of a sketchbook for those who are using it to study the development of Beethoven's musical thought. Thus it is important to examine the surviving sketchbooks in order to assess the degree of damage that each has suffered. And this examination may go on to suggest ways of repairing the damage; for once it is recognized that a book has lost leaves, it is a natural step to institute a search for them among the very large number of loose leaves that are in collections throughout the world. If missing leaves can be located there, it is possible to restore them, at any rate in conception, to their original position within the sketchbook. This may be called the work of reconstruction.

"Reconstructing Beethoven's Sketchbooks," an article by Douglas Johnson and myself that appeared in 1972, marked the first systematic attempt to consider these problems.[1] It described some of the techniques used in estimating the damage that a sketchbook had undergone—especially by means of an analysis of the book's physical makeup—and suggested methods of trying to reconstruct it. Further work since then has aimed at refining the techniques both of physical analysis and of reconstruction.[2] The discussion of the Leonore Sketchbook presented here may serve as a specimen of the kinds of problems that come up for consideration in this field, the arguments that are used, and the limits that must be accepted in the work. If it deals at considerable length with the description of the sketchbook furnished almost a century ago by Gustav Nottebohm, the reasons for this should become clear to the reader. And since it now appears plain that Nottebohm's conclusions about the date of the sketchbook were wrong, this seems an appropriate place, too, for a discussion of the chronology of the sketches for the opera *Leonore* and those of Mendelssohn 15 in particular. These problems need to be mastered before the sketchbook's main feature of interest, the nature and contents of the sketch material itself, can be judiciously and effectively studied.

Physical Descriptions of Mendelssohn 15

The sketchbook known as Mendelssohn 15 is today in the Staatsbibliothek Preussischer Kulturbesitz in Berlin. In the years immediately after Beethoven's death it was in the Artaria Collection. On 14 June 1834, however, it was sold (together with other autographs of Beethoven and Haydn) to Heinrich Beer of Berlin.[3] From Beer, the sketchbook eventually passed to Paul Mendelssohn and then to his son, Ernst von Mendelssohn-Bartholdy; on 26 June 1908, the latter presented it, together with the rest of the collection known as the "Paul und Ernst von Mendelssohn-Bartholdy'sche Stiftung", to the Royal Library in Berlin.

Mendelssohn 15 is the largest of the surviving sketchbooks, at any rate in its present form. It now has 173 oblong leaves (346 pages) of sixteen-stave paper. This paper is uniform throughout the whole book. Each leaf has a portion of the same watermark: a shield containing three six-pointed stars under a crown, and the

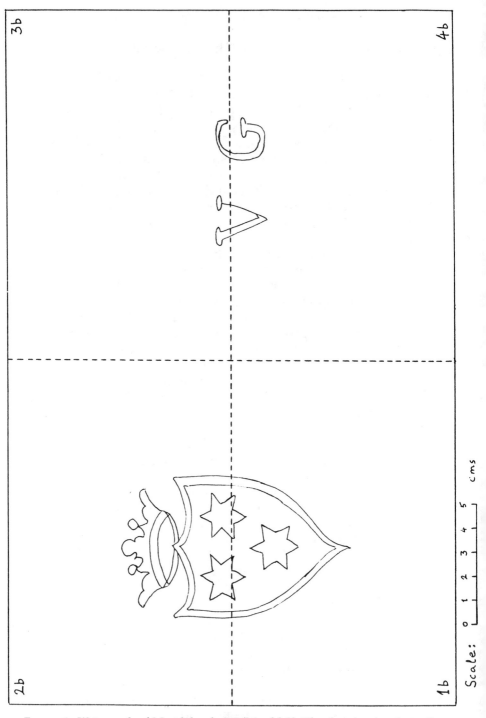

FIGURE 1. Watermark of Mendelssohn 15 (Mould B). The drawing is schematic. The watermark elements are to scale with each other, but are overly large in relation to the size of the sheet, indicated by the frame. For the sake of clarity, the chain lines have been omitted. The watermark of Mould A is very similar. But in Mould A the letter *V* is larger, and its right side as well as its left is formed by two lines.

letters *VG* (see figure 1). With a few exceptions (discussed below) the makeup of the book is also regular: it consists of single-sheet gatherings, that is, gatherings of two bifolia from a single sheet.[4] The sketchbook still has the brown half-cloth binding of Heinrich Beer's library.[5]

Unlike most sketchbooks, Mendelssohn 15 was examined and discussed by several scholars in the last century. The first, rather brief, account was that of Otto Jahn in 1863 (Jahn does not speak of a sketchbook but of "die reiche Sammlung von *Skizzenblättern zur ersten Leonore*, . . . jetzt in einen starken Band in Querfolio zusammengebunden").[6] Nine years later, Mendelssohn 15 was described more fully than any other sketchbook, in an appendix by A. W. Thayer to the second volume of his biography.[7] But the most extensive discussion is in Nottebohm's posthumously published essay in *Zweite Beethoveniana*.[8] It is necessary to take Nottebohm's scrutiny of Mendelssohn 15 as our starting point. His description is in some respects surprising, and it is important to decide whether this is the result of certain theories or preconceptions on his part, or whether (as in the case of several other sketchbooks that he described) he found Mendelssohn 15 in a somewhat different condition from the one that it is in today. Some of the subsequent discussion will be addressed to this problem.

After noting the number of the sketchbook's pages as 346 (the same number as is given by Thayer),[9] Nottebohm (pp. 409–10) described the book's physical condition as follows:

> Originally it consisted of two—or more precisely, of the second and third of four—sketchbooks connected by their content; the first of these, containing sketches for the first third of the opera, and the fourth of these, in which the work on the second finale (not completed here) and on the overture must have been continued, have been lost. When the book was bound some leaves were inserted in the wrong place [*verbunden*], and leaves were incorporated which do not belong here. With the leaves correctly bound, and with the exclusion of those that do not belong, the sequence of pages should run: pages 23–26; 1–22; 27–182; 187–198; 203–338. Some leaves are missing between page 26 and page 1. The leaves between page 182 and page 187, as well as those between page 198 and page 203, do not belong to the sketchbook. . . . The last four leaves too (pages 339–346) form no part of the sketchbook proper. . . . The sketchbook proper, beginning with page 23, therefore, and ending with page 338, . . . largely belongs to the year 1804.

Before we can evaluate this account, it is necessary to inquire into the date at which it was written and the probable extent of

Nottebohm's examination of sketchbooks. Only then shall we be in a position properly to compare his description with the book's present condition.

The Date of Nottebohm's Account

It is by no means clear when Nottebohm wrote his long essay on the sketchbook. The collection of Ernst von Mendelssohn-Bartholdy in Berlin contained two other sketchbooks besides Mendelssohn 15—those later know as Mendelssohn 6 and Mendelssohn 1. These came to the Royal Library with the rest of the collection in 1908; they were not to be found there at the end of World War II, and today they are in the Biblioteka Jagiellońska, Kraków. Unlike Mendelssohn 15, they were both described by Nottebohm in accounts published in the 1870s. The essay on Mendelssohn 6 first appeared in the *Musikalisches Wochenblatt* of 1875 (vol. 6, p. 413), and that on Mendelssohn 1 followed in the same journal in 1879 (vol. 10, p. 449), both being republished with slight revisions as chapters 33 and 34 of *Zweite Beethoveniana*. But there is evidence that as early as 1875, Nottebohm was familiar not only with other items in the Mendelssohn-Bartholdy collection but with Mendelssohn 15 itself, since at the start of one essay in the *Wochenblatt* of that year (vol. 6, p. 649), reviewing sketches for the Op. 59 quartets, he refers to the "sogenannte Leonore-Skizzenbuch" and to the fact that two leaves containing sketches for the F-major quartet (= pp. 183–86) had come to be wrongly bound up within the corpus of the sketchbook. This observation suggests that its content, and perhaps its makeup or physical structure, had already been examined rather closely by then.[10]

Nottebohm's long series of contributions to the *Wochenblatt* under the running title of "Neue Beethoveniana" finally ceased in October 1879 without any further reference to Mendelssohn 15 having appeared. Yet it is impossible to believe that the essay on Mendelssohn 15 had not been written (at any rate partly) by then. For its opening paragraph includes the assertion that another sketchbook containing sketches for the first third of the opera had been lost. Now such a remark could not have been made by Nottebohm after the year 1879 (at latest), since he must by then have become familiar with the contents of the "Eroica" Sketchbook, Landsberg 6; Nottebohm's monograph on Landsberg 6, which briefly describes the sketches for the first five numbers of the opera that are found on pages 146 to 171 of that sketchbook, was published in July 1880 and was the last piece of writing that he saw through the press before his death in 1882.[11] It may even be that the discovery of the Act I sketches for *Leonore* in Landsberg 6 led Nottebohm to postpone the publication of an already completed essay on Mendelssohn 15, the length of which would in any case have suggested presentation in a monograph rather than in a further series of "Neue Beethoveniana."

The important points, then, are that the essay on Mendelssohn 15 must have been completed by 1879 at the latest, and that it cannot represent Nottebohm's final opinions on certain aspects of the sketchbook's contents.[12]

Nottebohm's Examination of Sketchbooks

Although it was not unusual for Nottebohm to comment on the condition of a sketchbook and, in particular, on the places at which it had lost some of its leaves, he left few clues as to his method of scrutiny. Most of these are contained in a single essay, that on the Petter Sketchbook, published in the *Musikalisches Wochenblatt* in 1879 (vol. 10, pp. 193–95, 205–6, 213–14, 229–30). This essay was professedly a polemic against certain passages in the third volume of Thayer's biography, which had come out earlier in the same year and had assigned a very different chronology to the contents of the Petter Sketchbook.[13]

Nottebohm's description of his methods (which was omitted from the version of the essay subsequently printed as chapter 31 of *Zweite Beethoveniana*) is interesting both for what it includes and for what it omits:

> I compared the paper of the individual leaves, I compared the staves, and I studied the leaves at the fold, at the place at which they were gathered together by the binder, in order to establish which leaves were connected at the fold and which were not. Such a scrutiny, combined with a comparison of the sketches from the point of view of the connections between them, was essential for determining the original sequence of leaves (precisely or approximately), for disclosing possible gaps, and for detecting any leaves that had been wrongly introduced, whether through oversight or from any other circumstance.

This enabled the makeup of the Petter Sketchbook to be somewhat loosely tabulated:

The contents of the book are arranged here in accordance with the sections put together by the binder. The leaves . . . are arranged according to whether or not they fit together at the fold so as to make a bifolium if they are a pair or a gathering if they are a foursome.

From this account of the methods that he used in reconstructing Petter (one of the last sketchbooks to be described in the "Neue Beethoveniana" series in the *Musikalisches Wochenblatt*), we may conclude that Nottebohm was interested in evidence of how a volume was stitched together ("sewn"), and that he tried to match leaves in twos and fours. He also paid great attention to sketch content and to the sequence of sketches. But he does not seem to have examined watermarks.[14] Nor, apparently, did he conceive of the *sheet* as the unit of construction when he examined Petter, for he was not puzzled or dismayed if only one of the two bifolia from a sheet was present. Furthermore, he seems to have expressed no views on the likely overall size of an individual volume.

Evaluation of Nottebohm's Description of Mendelssohn 15

Let us begin with Nottebohm's statement that Mendelssohn 15 "originally consisted of two sketchbooks connected by their content." Why should he have said that? He presents no evidence on the point. And today there is no obvious way in which the sketchbook could be said to fall into two (presumably approximately equal) sections. It is tightly bound within a single unrevealing cover, and is without any overt signs of disunity. Is it possible, then, that in the 1870s the sketchbook still betrayed signs of a binary origin?

In some respects at least it may have presented a somewhat different appearance in the past. For even though it is still contained within the cover supplied by Heinrich Beer, the sketchbook has been rebound more than once: in about 1930, for example, and again in 1970.[15] It is possible therefore that between the time that it was examined by Nottebohm and the present day some piece of evidence—wrappers, for instance, or at least the remnant of a wrapper somewhere in the middle of the sketchbook—was removed. It may even be significant that, as pointed out above, Jahn did not speak of a sketchbook at all but only of a "reiche Sammlung von Skizzenblättern"—a phrase that could suggest that the leaves were much less firmly bound than they are at present.[16]

The first explicit statement that Mendelssohn 15 once consisted of two sketchbooks was made by Thayer in 1872. In his appendix on the sketchbook (vol. 2, p. 393) he wrote: "Only in respect of its contents, its cover, and its large dimensions does the sketchbook differ from many others; and, as is stated in the text, this last distinction is doubtless due to the fact that it consists of the second and third of three sketchbooks, which have been bound together." Unfortunately one searches "the text" in vain for any further clarification of this last point. All that can be found is a comment on page 279 that the sketchbook begins in the middle of the Prisoners' Chorus, and that any sketches for the preceding numbers take the form of revisions of work already sketched elsewhere. Like Nottebohm, Thayer nowhere brings forward evidence for the double origin of the sketchbook.

When one compares Nottebohm's description of Mendelssohn 15 in this respect with Thayer's, the similarities are at once apparent. For Thayer, it was the second and third of three sketchbooks used for the opera; for Nottebohm, it was the second and third of four. It looks as though Nottebohm is emending Thayer, though (as usual) he does not refer to him directly. And if we ask what could have led both Thayer and Nottebohm to state that Mendelssohn 15 was originally two sketchbooks—accepting for a moment the absence of any physical features of the book visible in their time and not in ours—it seems we can get no further than some assumption on their part concerning the "normal" length of a Beethoven sketchbook. They found Mendelssohn 15 too big, and decided that it must be two sketchbooks. (Neither scholar refers at any point in discussing the sketches to the transition from one book to the other.) The evidence available today confirms that Beethoven worked with sketchbooks of more than one length; there was no maximum length, but certain dimensions were evidently customary. Several of the sketchbooks that Beethoven used in the years between 1801 and 1808 originally had 96 leaves, just about half the number of leaves in Mendelssohn 15. To Thayer and to Nottebohm (who devoted his monographs to describing two of them), these no doubt were the books of "normal" length. But other books, such as Grasnick 1 (1798–99), Grasnick 2 (1799), and Grasnick 3 (1808) seem originally to have had 48 leaves. There is no compelling reason, therefore, why Beethoven should not have opted for a particularly large volume—one with 192 (= 2 × 96) leaves, for example—in getting to grips with the longest work that he was ever to write. There is an easy explanation for the fact that the number of leaves in many of these sketchbooks is a multiple of 24: music paper was normally sold in "Bücher" of 24 *Bogen*, that is, 24 large sheets of paper (making 96 leaves). In fact, a "Buch" of music paper would be exactly the right quantity to make one of Thayer's and Nottebohm's "normal" sketchbooks.

Apart from his claim that the sketchbook was once two sketchbooks, Nottebohm suggests a precise sequence of leaves: pages 23–26, a gap, 1–22, 27–182, 187–98, 203–338. Once again he does not present his evidence, and the possibility must remain open that he saw features in the sketchbook that are no longer present. But the suspicion also arises that the almost unbroken structure of Mendelssohn 15 forced Nottebohm to notice that the gatherings had originally been ones of four leaves each, and to conclude that where a gathering consisted of no more than two leaves it was either defective or an interpolation. In this way the bifolia pages 23–26, 183–86, and 199–202 would come under special scrutiny, and the last few leaves of the book would also be suspect. It may be that the much greater damage that the Petter Sketchbook had undergone prevented him from perceiving the same underlying single-sheet gathering structure of that book.

Page 1 of Mendelssohn 15 still carries Artaria's seal and inscription ("Artaria und Comp / Wien, 14 Juny 1834"). By that date, the initial leaves must already have been rearranged. For Nottebohm's claim that pages 23–26 are misplaced and should precede page 1 is perfectly correct, although his further statement that leaves have been lost between page 26 and page 1 can be shown to be false. Let us try to get inside Nottebohm's mind here. It was probably the fact that pages 23–26 were not a full gathering of four leaves but only a pair of leaves that first led him to suspect that they could be misplaced, and that other leaves had been lost at this point. The contents of page 23, sketches for the tenth number of the opera (the duet "Um in der Ehe froh zu leben"), and of pages 24–26, very early sketches for the Prisoners' Chorus, will then have led him to place this bifolium at the beginning of the book. Or he may have noticed that a sketch for the same chorus on page 22, staves 12/13, is continued directly on page 27, staves 6/7, exposing pages 23–26 as an interpolation. There is also an offset onto page 22, stave 2, from page 27, stave 2.

Today we are in a position to correct Nottebohm's account a little. It is now clear that page 1 followed page 26 directly. An inkblot proves this: a blotted note on stave 4 of page 1 has produced an offset on stave 4 of page 26. No leaves therefore were lost between page 26 and page 1; but instead it can be shown that two leaves were lost between pages 24 and 25. These two leaves, a bifolium, are now in one of the sketch miscellanies that was once owned by Ludwig Landsberg and that derived ultimately from the Artaria Collection: Landsberg 12, pages 43–46. The bifolium has stitch holes along the midline fold. Accordingly the correct sequence of leaves at the beginning of our sketchbook is: Mendelssohn 15, pages 23–24; Landsberg 12, pages 43–46; Mendelssohn 15, pages 25–26; 1–22, 27 ff.

Nottebohm's assertion that pages 183–86 and 199–202 are interpolations is also of considerable interest and is probably justified. Pages 183–86 contain sketches for the last three movements of the quartet Op. 59, no. 1. They are discussed in Nottebohm's essay on the Op. 59 quartets in the *Musikalisches Wochenblatt* of 1875 (vol. 6, p. 649; reprinted with very small changes in *Zweite Beethoveniana*, chap. 11). This bifolium, which comes in the middle of sketches for the first movement of the "Appassionata" Sonata, Op. 57, clearly does not belong in its present position and has evidently been wrongly introduced there. Though he does not say so, Nottebohm could well have observed that inkblots on page 182 have produced offsets on page 187; this proves that pages 183–86 cannot originally have stood there. And he may have decided that sketches for the later movements of Op. 59, no. 1 (the autograph of which is inscribed "Quartetto angefangen am 26ten Maj—1806") could not be fitted into the time span of the sketchbook (on this point see below).

Pages 199–202, a bifolium with sketches for the Triple Concerto, Op. 56, also do not seem to make much sense in their present position; they too interrupt a set of sketches for the first movement of Op. 57. But here it is less easy to prove that they are misplaced, since the leaf after pages 197–98 has been lost, and—supposing for a moment pages 199–202 to be *correctly* placed—we should in that case have to assume that there had once been a further bifolium, now lost, enclosing pages 199–202. Thus the jarring transition from page 198 to page 199 could have been softened by the contents of two (lost) leaves. So it seems probable that Nottebohm's claim that pages 199–202 are misplaced rested on other grounds: either on some physical feature of this part of the sketchbook visible to him but denied to us, or to the fact that the sketch content of these pages seems to be "earlier" than that of certain pages found nearer the beginning of the sketchbook: pages 140–42, for example. The theme of the second movement of

Op. 56 is sketched, for instance, with four introductory bars on pages 14 and 15, and they are hinted at again on pages 199–200. But these introductory bars do not occur in the final version of the movement, and have (it seems) already been dropped on page 140. If pages 199–202 are indeed correctly placed, this would represent a regression. Such a line of thought, though it begs a number of questions concerning Beethoven's way of thinking and working and even the manner and order in which he used up the pages of a sketchbook, would at any rate be highly characteristic of Nottebohm.

Both these probably intrusive bifolia, it need hardly be said, have the same watermark and rastrology as the unexceptionable parts of Mendelssohn 15; and the same is true of the last four leaves (pp. 339–46) that according to Nottebohm "form no part of the sketchbook proper." They are evidently single leaves, tacked onto the end of the book in no kind of structure. In the 1870s, they may have been inserted only loosely at the end of the book; Thayer (vol. 2, p. 400) described pages 341–46 as "diese verkehrt gebundenen Seiten," although today they appear to be bound in the correct way round. Nottebohm recognized that their content consisted in part of sketches that antedate the first 1805 performance of the opera, and in part of superimposed sketches made between the 1805 and 1806 productions;[17] some of them—for instance, the sketches for Op. 56 on pages 341–42—seem to be as early as any in Mendelssohn 15. He was surely right in rejecting them as a part of the book.

Reconstruction of Mendelssohn 15

The reconstruction that is offered here takes advantage of the various techniques that are at present available for such a task: analysis of the physical makeup and determination of the original gathering structure, study of the continuity and sequence of sketch contents, investigation of inkblots and their offsets, assessment of the descriptions given by earlier (mainly nineteenth-century) scholars.

In all but one aspect the reconstruction presents no problem. For the sketchbook has been well preserved. Essentially it consists of forty-four four-leaf gatherings. Of these thirty-four are still intact; eight have lost one leaf, and two have lost two leaves. (A forty-fifth gathering probably begins with pp. 337/338, but this has lost its last three leaves.) Two of the lost leaves are identified as

Landsberg 12, pages 43–46; they can be mentally restored to their original position. The reconstruction follows Nottebohm in rejecting the bifolia pages 183–86 and pages 199–202 as interpolations; the four last leaves (pp. 339–46) are also excluded. The correctness of the order in which the gatherings stand today—apart from pages 23–26—is confirmed in general not only by the sequence of the musical content but also by inkblot links; these are indicated on the diagram of the sketchbook's makeup.

But the problem of whether it was originally one sketchbook or two still remains. Today, as we have seen, there is no evidence to prove that Mendelssohn 15 was once two sketchbooks of ninety-six leaves each. Even the stitching of the gatherings is spaced at the same sort of intervals from top to bottom of gatherings both at the beginning and at the end of the sketchbook.[18] If it was formed out of two books, then, they must have been remarkably uniform (though perhaps no more so than the sketchbooks Grasnick 1 and Grasnick 2). Only the confident assertions of Thayer and Nottebohm that Mendelssohn 15 was once two books must give us pause, especially when we find that, within its guileless cover, some of the leaves have been rearranged and other leaves wrongly inserted.

Since there is no appreciable gap in the sequence of sketches around the middle of the book, does it make any difference if we assume that it was formed out of two books? The answer is that such an assumption has certain implications for the work of reconstruction. This becomes clearer when one begins to consider at what point the two books should be separated. For certain conditions need to be observed. The division should not be more than ninety-six leaves from the beginning or more than ninety-six leaves from the end. And it must not separate two gatherings that are clearly linked by inkblots from Beethoven's entries.

In practice, these conditions allow for very few possibilities—they indicate that any break can only have been after one of the following pages: 164, 168, 174, 180. And even those choices can be narrowed down by further considerations. The first of them is perhaps unlikely, since a phrase begun at the bottom of page 164 is continued near the top of page 165. The last is called into question by the evidence of a marginal blot on the outside edge (fore edge) of several leaves, opposite staves 12 and 13. This blot links pages 179–82 and 187–98; significantly, it skips over the two interpolated leaves, pages 183–86. If the blot dates from Beethoven's time, as is likely, it eliminates page 180 from being the last page of the first book. Thus we are left with a choice of page 168 and page 174. The

BOOK 1

I				XII			
	23/24	3b			85/86	2b	
	L 12, 43/44	4b			87/88	1b	
	L 12, 45/46	1b			89/90	4b	
	25/26	2b	*		91/92	3b	* *

II				XIII			
	1/2	1a			93/94	3a	
	3/4	2a			95/96	4a	
	5/6	3a			97/98	1a	
	7/8	4a	+		99/100	2a	*

III				XIV			
	9/10	1b			101/102	2a	
	11/12	2b			103/104	1a	
	13/14	3b			105/106	4a	
	15/16	4b			107/108	3a	*

IV				XV			
	17/18	1a			109/110	1b	
	19/20	2a			111/112	2b	
	21/22	3a	* +		113/114	3b	
	27/28	4a	* +		115/116	4b	*

V				XVI			
	29/30	2b			117/118	2b	
	31/32	1b			119/120	1b	
	33/34	4b			121/122	4b	
	35/36	3b	*		123/124	3b	* +

VI				XVII			
	37/38	4b			125/126	3b	
	39/40	3b			127/128	4b	
	41/42	2b			129/130	1b	
	43/44	1b	*		131/132	2b	*

VII				XVIII			
	45/46	1b			133/134	1b	
	47/48	2b			135/136	2b	
	49/50	3b			137/138	3b	
	51/52	4b			139/140	4b	+

VIII				XIX			
	53/54	2b			141/142	2a	
	55/56	1b			143/144	1a	
	57/58	4b			145/146	4a	
	59/60	3b	*		147/148	3a	+

IX				XX			
	61/62	3b			149/150	3a	
	63/64	4b			151/152	4a	
	65/66	1b			153/154	1a	
	67/68	2b			155/156	2a	*

X				XXI			
	69/70	4b			157/158	2a	
	71/72	3b			159/160	1a	
	73/74	2b			161/162	4a	
	75/76	1b	+		163/164	3a	+

XI				XXII			
	77/78	3b			165/166	2b	
	79/80	4b			167/168	1b	
	81/82	1b			[A (stub)]	[4b]	
	83/84	2b			[B (stub)]	[3b]	

FIGURE 2. Reconstruction of Mendelssohn 15. The diagram on these two pages shows a hypothetical reconstruction of Mendelssohn 15. It assumes that the sketchbook originally consisted of two separate books, each of ninety-six leaves, and that leaves were subsequently lost—eight leaves at the beginning of Book 1, and seven leaves at the end of Book 2. Ten further leaves, lettered here from [A] to [J], are today missing at various places within Mendelssohn 15.

BOOK 2

Sheet	Pages	Watermark
I	[C]	[1b]
	169/170	2b
	171/172	3b
	173/174	4b
II	175/176	1a
	177/178	2a
	179/180	3a
	[D (stub)]	[4a]
III	181/182 *	2b
	187/188	1b
	189/190	4b
	191/192	3b
* +		
IV	193/194	2b
	195/196	1b
	197/198	4b
	[E]	[3b]
V	203/204	3b
	205/206	4b
	207/208	1b
	209/210	2b
*		
VI	211/212	1b
	213/214	2b
	215/216	3b
	217/218	4b
*		
VII	219/220	1b
	221/222	2b
	223/224	3b
	225/226	4b
VIII	227/228	3b
	[F]	[4b]
	229/230	1b
	231/232	2b
*		
IX	233/234	3b
	235/236	4b
	237/238	1b
	239/240	2b
*		
X	241/242	4b
	243/244	3b
	245/246	2b
	247/248	1b
*		
XI	249/250	1b
	251/252	2b
	253/254	3b
	255/256	4b
*		
XII	257/258	3a
	259/260	4a
	261/262	1a
	263/264	2a
*		

Sheet	Pages	Watermark
XIII	265/266	1b
	267/268	2b
	269/270	3b
	271/272	4b
*		
XIV	273/274	1a
	275/276	2a
	277/278	3a
	279/280	4a
XV	281/282	3a
	283/284	4a
	285/286	1a
	287/288	2a
* +		
XVI	289/290	3b
	291/292	4b
	293/294	1b
	295/296	2b
+		
XVII	297/298	3b
	[G]	[4b]
	299/300	1b
	301/302	2b
*		
XVIII	303/304	3b
	305/306	4b
	307/308	1b
	[H (stub)]	[2b]
XIX	309/310	2b
	311/312	1b
	313/314	4b
	315/316	3b
+		
XX	317/318	2b
	319/320	1b
	321/322	4b
	323/324	3b
XXI	[I (stub)]	[1a]
	325/326	2a
	327/328	3a
	329/330	4a
XXII	331/332	2b
	[J]	[1b]
	333/334	4b
	335/336	3b
+		
XXIII	337/338	3b
Not part of the sketch-book	339/340	1b
	341/342	2b
	343/344	1b
	345/346	4?
Inter-polated leaves	183/184	1b
	185/186	4b
	199/200	4b
	201/202	1b

* Link provided by inkblot and offset
+ Link provided by musical continuity

The information used in the reconstruction is reproduced schematically in the three columns of the diagram:

The first column shows the original sheets by means of Roman numerals.

The second column shows the original gathering structure and the present page numbering. The position of leaves no longer present is shown by letters in square brackets.

Two further signs are used between sheets, to indicate that they are still in their original order. A star marks the connection between two sheets that is demonstrated by inkblots. A cross indicates an unambiguous musical continuity from the last page of one sheet to the first page of the next. (These signs are not used within a sheet, unless there is a gap in the page numbering.)

The third column shows the watermark quadrant of each leaf, together with its mould ("a" or "b").

former is much the more convincing; not only does it make the more decisive break in the sketch contents (it is itself blank), but it is followed by a gap where three leaves are missing, two from its own gathering, and one from the next. It is easy to imagine that leaves could have become detached at the very end (and beginning) of a sketchbook. If there were once two books, then, it is likely that the first ended with the two leaves that are now missing after page 168 and that the second began with the leaf, also lost, before page 169. But nothing seems to have been lost at this point beyond the three leaves just mentioned.

The consequences of this division can be seen from the diagram (figure 2). In spite of the lack of evidence on this point, we have provisionally divided Mendelssohn 15 between pages 168 and 169. Book 1 accordingly consists of twenty-two gatherings, and we must presume that two more gatherings that once preceded the first gathering (itself disrupted) have been lost. The two last leaves of Book 1, [A] and [B], have also vanished. Book 2 consists of twenty-two gatherings, too, and probably has the first leaf of the twenty-third as well. Thus seven leaves (1¾ gatherings) have been lost at the end, and eight others (leaves [C] to [J]) before that. The theoretical division of Mendelssohn 15 into two books results in a more exact localization of the missing gatherings. It is unfortunate, therefore, that it should still not be clear whether such a division is justified.

Date of the Sketchbook

If we assume that not more than a few leaves have been lost from the beginning of Mendelssohn 15, and that the sketchbook proper ends at about page 338, we have still to determine the period within which the book was used. The later limit is easy enough to fix. Page 338 includes early sketches for the overture *Leonore No. 2*, written for the first production of the opera and played at the first performance on 20 November 1805.[19] Since this performance had been delayed for over a month by censorship difficulties, it is likely that the overture was completed by the end of October 1805. We can take that month, then, as the effective *terminus ante quem* for the final pages of Mendelssohn 15.

But the date at which Beethoven started to use the sketchbook is less easy to determine with precision, and involves us in a general discussion of the opera's chronology and an examination of the sketches for it that lie outside Mendelssohn 15. There are not in fact so many of these. The earliest of them are the sketches for some of the numbers of Act 1 that are to be found in Landberg 6, pages 146–57 and 160–71. And the same sketchbook also contains sketches for the operatic venture that almost immediately preceded the first sketches for *Leonore* (from which it is separated only by the sketches for the "Waldstein" Sonata): the work on Schikaneder's libretto *Vestas Feuer,* which Beethoven abandoned after only a short while. A chronological framework for these two sets of sketches in Landsberg 6 is provided by five contemporary letters, only the second of which was known to Nottebohm:

1. 22 October 1803. Ferdinand Ries to Simrock: "Beethoven will soon receive the subject of his opera" (for the Theater-an-der-Wien).[20]

2. 2 November 1803. Beethoven to Alexander Macco in Prague: "I am only now beginning to work on my opera."

3. 12 November 1803. Georg August Griesinger to Breitkopf & Härtel: "At present he [Beethoven] is composing an opera by Schikaneder, but he told me himself that he is looking out for reasonable texts."[21]

4. 4 January 1804. Griesinger to Breitkopf & Härtel: "Beethoven has recently given Schikaneder his opera back because he feels the text is much too ungrateful."[22]

5. 4 January 1804. Beethoven to Rochlitz: "I have completely broken with Schikaneder. . . . I have now quickly had an old French libretto adapted [J. N. Bouilly's *Léonore ou l'amour conjugal*] and am now beginning to work on it."

Although there are certain risks in taking Beethoven's statements in correspondence or in conversation too literally in matters of chronology, these five letters paint a consistent picture. They indicate that Beethoven worked on *Vestas Feuer* from the end of October or the beginning of November 1803 till near the end of the year, before he finally abandoned it, and that he then began sketching *Leonore* in the first months of 1804. This dating seems to be further confirmed by the sketches on pages 172–79 of Landsberg 6 that immediately follow those for *Leonore.* They are for the revision of the first and second numbers in the oratorio *Christus am Oelberge,* Op. 85, which had received its first performance in April 1803. Since the oratorio was given again in Vienna on 27 March 1804, this probably provided Beethoven with an opportunity and incentive for revising the work.

The Landsberg 6 sketches for *Leonore,* then, almost certainly fall into the period of January-March 1804.[23] They are the earliest to have come down to us, and they are confined to the first five numbers of the opera; the sixth and last number of the first act, the terzetto "Gut, Söhnchen, gut," is not represented there. Apart from a group of four leaves in the Berlin miscellany Landsberg 10 (SV 64), pages 21–28, with sketches for the *Leonore No. 2* overture, the only other sketchleaves for the opera to have survived, apart from Mendelssohn 15, are a group of ten, which are now in four different locations. They are all somewhat later in date than those in Landsberg 6, but they are still restricted to the first act of the opera. Since they are linked not only by identity of paper type but also to some extent by content, they may be taken together:

    (a) Berlin, SPK, aut. 19e, fols. 95–98: 4 leaves.
    (b) Bonn, Beethovenhaus, BSk 17/65a: 3 leaves.[24]
    (c) Basel, Georg Floersheim: 1 leaf.
    (d) Vienna, GdM, A 39: 2 leaves.

All these sketchleaves have sixteen staves to the page and their paper has the same watermark: the letters *VB* under a crown (or baldachin) and three crescent moons, of which the largest has a face in profile. Several of the leaves have been cut or torn somewhat crudely along the upper margin, leaving an irregular edge. To judge from their appearance, none of them formed part of a sketchbook, and they were no doubt used by Beethoven in the form of loose bifolia or gathered pairs of bifolia.

There are obvious links at any rate between (a), (b), and (c). Some of the contents of (a) have been extremely well-known since the time of Nottebohm, and others were recently discussed by Virneisel.[25] These four leaves contain: piano exercises; early sketches for the Fifth Symphony, Op. 67, and the Fourth Piano Concerto, Op. 58 (with the theme subsequently used for the Prisoners' Chorus in *Leonore* here used in G major for the finale of the Concerto); transcriptions of fugues by Georg Muffat; a passage near the end of the sixth number of the first act of *Leonore* ("Gut, Söhnchen, gut"); a phrase from the terzetto (no. 6) of *Christus am Oelberge*; sketches for two sonnets by Petrarch in the translation of A. F. K. Streckfuss; and four Scottish melodies, no doubt supplied to Beethoven by George Thompson of Edinburgh.

The four leaves of (a) are from the same sheet of paper; and the same is true of the three leaves of (b) and the single leaf (c).[26] The contents of (b) and (c) include: transcriptions of passages from Mozart's *Zauberflöte*, act 1, no. 5, and from Cherubini's *Les deux journées* (known in German-speaking countries as *Der Wasserträger, Graf Armand,* or *Die Tage der Gefahr*), act 1, no. 3 (terzetto) and finale; sketches for the second number of *Leonore* ("Jetzt, Schätzchen"); and two very early ideas for the finale of the Fourth Symphony, Op. 60. But what links (b) and (c) to (a) is the presence on the first page of (b) of the same Scottish melody that is to be found at the bottom of the last page of (a). This connection is striking enough for us to be able to regard (a), (b), and (c) as a sequence of eight leaves. Finally, (d) consists of two separate leaves that were probably originally a bifolium, with sketches for Matthisson's *Wunsch* and for "Gut, Söhnchen, gut." The sketches for this last number are quite close to the version finally adopted, but since they concern the first part of the terzetto only, their exact relation to (a), which contains sketches only for its final bars, cannot be determined.[27]

How are these ten leaves to be dated? The best clues are provided by (a). The phrase from Op. 85 is connected with a passage in the oratorio's sixth number that caused Beethoven dissatisfaction and led to small changes being made. The possibility cannot be excluded that these changes were made about March 1804, at the same time as the revisions to the first two numbers of the oratorio, the sketches for which are in Landsberg 6. But in the *Stichvorlage* for the first edition of the oratorio (British Library, London, Egerton 2727), all the changes in the first two numbers are supplied by a copyist; those in the sixth number are in Beethoven's own hand. It seems more likely that they were further revisions made after the March performance of the oratorio but before the score of the work was offered to Breitkopf & Härtel in August 1804.

More help can be gained from the settings of the Petrarch sonnets. Streckfuss's *Gedichte,* from which the translations were taken, was announced by their publisher Degen in the *Wiener Zeitung* on 24 May 1804. Provided that Beethoven did not see an advance copy of the book, and provided that the publisher's announcement was not unduly delayed (Degen's previous advertisement had appeared on 7 April 1804), it can be accepted that the last page of (a) was not written until after 24 May 1804. Even if one or other of those possibilities is allowed for, it still seems unlikely that these ten leaves would be sketched before about May or June of 1804. And from this it follows that Mendelssohn 15, which passes over the first act of the opera[28] and (except for its first surviving page, p. 23) starts with the finale of the second act, can scarcely have been begun before June or July 1804. The date may even be a little later, for it was presumably before beginning Mendelssohn 15 that Beethoven sketched the first four vocal numbers of the second act. But those sketchleaves have not survived. The *outside* limits for the complete sketchbook, then, would be May or June 1804 and October 1805.

The Date on Page 291 of Mendelssohn 15

As the title of his essay shows ("Ein Skizzenbuch aus dem Jahre 1804"), Nottebohm was determined to limit the contents of the sketchbook to the year 1804. Although the evidence presented in the previous section indicates that Beethoven used the sketchbook over about eighteen months from the summer of 1804 to October 1805, we cannot leave the matter without a brief examination of an entry on page 291 of the sketchbook, since it formed the subject of a strenuous discussion by Nottebohm. This entry runs:

> am 2ten Juni—finale imer simpler—alle Klawier-Musick ebenfalls— Gott weiß es—warum auf mich noch meine Klawier-Musick imer den schlechtesten eindruck [macht,] besonders wenn sie schlecht gespielt wird.

> (On 2 June—finale always simpler—the same goes for all piano music. God knows why my piano music still always makes the poorest impression on me, especially when it is badly played.)

Was this written on 2 June 1804 or 2 June 1805? Nottebohm, who must have noted that Thayer (vol. 2, p. 278) opted by implication for the latter date, argues vigorously for the former in a long footnote that must be quoted in full (p. 446n):

> Our assumption is that this comment falls in the year 1804. The assumption advanced elsewhere that the comment was written in 1805 is contradicted in the first place by the shortness of the time remaining to Beethoven for completing the opera. The overture had not yet been started, work on the second finale had hardly advanced beyond its first stage, and other numbers in the second act [Nottebohm means the third act of the original "1805" version] had not been finished when the comment was written; and it is improbable that those pieces were got ready or that the parts could have been written out, the necessary rehearsals held, etc., in the short time from 2 June 1805 to the day of the first performance (20 November 1805). It can moreover be safely assumed that the opera was finished a fairly long time before the first performance. This is clear from the following reports. The Vienna

correspondent of the Leipzig *Allgemeine musikalische Zeitung* for January 1806 (p. 237) says in his report on the first performance: "The most notable of the musical productions was no doubt the long-awaited opera by Beethoven." Ferdinand Ries relates (*Biographische Notizen*, p. 102): "One day when a small gathering of persons, which included Beethoven and myself, had breakfasted with the Prince (Lichnowsky) after the concert in the Augarten, it was decided to go to Beethoven's house in order to hear his opera *Leonore* which had not yet been performed." What Ries relates can only of course have happened in 1805 and not before.

It will be seen that Nottebohm's arguments are not in fact strong. There is a mass of evidence to show that Beethoven composed rapidly, especially when under pressure or to meet a deadline. The work still to be done does not seem excessive, and may have been more advanced than Nottebohm's account implies. The second finale, for instance, had already been sketched from page 244—that is to say, on every one of the forty-seven pages that precede the "2 June" entry on page 291. The difficulty in having the parts copied out (once the music had been written) is an unreal one; in any case there is some evidence (based on watermarks) that each number of the opera was written out in an *Abschrift* by Beethoven's chief copyist of the time as soon as it was composed.[29] And rehearsals need not always have been rehearsals of the complete opera. The quotation from Ries proves only that there was a private performance of some part of the opera no later than September 1805 (when Ries had to leave Vienna in order to report for military service at Bonn), and the comment of the *Allgemeine musikalische Zeitung* proves only that the opera on which Vienna's leading composer had been working since the beginning of 1804 was "long-awaited."

Today it is overwhelmingly clear that Nottebohm was mistaken, and that the entry on page 291 was made on 2 June 1805. This date, it will be seen, accords well with our overall chronology for Mendelssohn 15. And an additional piece of confirmation comes from a further source that was unknown to Nottebohm: a letter from Countess Josephine Deym to her mother, dated 24 March 1805 and first published by La Mara in 1920.[30] The letter refers to the song "An die Hoffnung," Op. 32, which was eventually published in September 1805:

Der gute Beethoven hat mir ein hübsches Lied, das er auf einen Text aus der Urania "an die Hoffnung" für mich geschrieben, zum Geschenk gemacht.

(Dear Beethoven has made me a present of a lovely song, "An die Hoffnung," which he wrote for me to words from [Tiedge's] *Urania*.)

What is almost certainly the same song ("un air pour Pepi"—one of Josephine's nicknames) is referred to in an undated letter (early January 1805?) from Charlotte von Brunsvik to her sister Therese, and in Therese's letters of 17 and 20 January 1805 to her brother Franz and to Charlotte; it may possibly have been a New Year's present from Beethoven to Josephine.[31] Sketches from Op. 32 are found on pages 151–57 of Mendelssohn 15; the evidence of the letters indicates that that part of the sketchbook was being used by Beethoven about December 1804—a date consistent with a date of 2 June 1805 for page 291, but quite inconsistent with Nottebohm's attempt to restrict the sketchbook proper to the year 1804.

Notes

This essay first appeared (in a German translation by Sieghard Brandenburg) in the *Beethoven-Jahrbuch* 9 (1977). The English text, now printed for the first time, has been revised in a few passages to bring it up to date but is otherwise unchanged.

1. Douglas Johnson and Alan Tyson, "Reconstructing Beethoven's Sketchbooks," *Journal of the American Musicological Society* 25 (1972): 137–56.

2. The most important of these refinements is the use of the distinction between the two ("twin") moulds of the watermark. See especially Alan Tyson, "A Reconstruction of the Pastoral Symphony Sketchbook (British Museum Add. MS 31766)," in *Beethoven Studies* 1, ed. Alan Tyson (New York, 1973), pp. 67–96, and (for an overview of the rules for the description of watermarks) the appendix to Alan Tyson, "The Problem of Beethoven's 'First' *Leonore* Overture," *Journal of the American Musicological Society* 28 (1975): 332–34.

3. See Douglas Johnson, "The Artaria Collection of Beethoven Manuscripts: A New Source," *Beethoven Studies* 1, pp. 235–36 and footnote 93.

4. The handmade paper used in Beethoven's sketchbooks was originally manufactured in sheets measuring approximately 40–50 × 60–70 cm (vertical dimensions first). These sheets, when folded and cut, formed two bifolia, one of which was gathered inside the other. In such a case, we can speak of "single-sheet gatherings."

5. For a brief description of the sketchbook's present condition, as well as a very exact stave-by-stave itemization of its sketch contents, the reader is referred to the long section (pp. 231–77) devoted to Mendelssohn 15 in Hans-Günter Klein's catalogue of the Beethoven collection in the Staatsbibliothek Preussischer Kulturbesitz: *Ludwig van Beethoven: :Autographe und Abschriften* (Berlin, 1975).

6. Otto Jahn, "Leonore oder Fidelio," *Allgemeine musikalische Zeitung* (1863), cols. 381–85 (27 May) and 397–401, with *Beilage* (3 June); reprinted in Jahn's *Gesammelte Aufsätze über Musik* (Leipzig, 1866), p. 242ff.

7. A. W. Thayer, *Ludwig van Beethovens Leben*, vol. 2 (Berlin, 1872), pp. 393–400. It is clear from a remark on page 403 that Thayer had examined the book in

1870. The sketchbook is also discussed in the main text of the volume, pp. 278–81.

8. Gustav Nottebohm, *Zweite Beethoveniana: nachgelassene Aufsätze* (Leipzig, 1887), chap. 44, pp. 409–59.

9. Thayer carelessly writes of "176" leaves, but he cites the pages present as "346."

10. The start of this essay was reprinted with minimal alterations at the beginning of chapter 11 of *Zweite Beethoveniana*.

11. Gustav Nottebohm, *Ein Skizzenbuch von Beethoven aus dem Jahre 1803* (Leipzig, 1880). The book was advertised in the *Musikalisches Wochenblatt* on 30 July 1880.

12. It is possible that it was written much earlier. In October 1874 Brahms wrote to Fritz Simrock in Berlin: "Nottebohm has written accounts of two very interesting sketchbooks (like the one published by Härtel [i.e., Nottebohm's 1865 monograph on the Kessler Sketchbook]). Wouldn't you be interested in publishing them?" Brahms, *Brief-Wechsel* 9 (1917): 184. Some of the contents of Landsberg 6, including the sketches for the beginning of *Leonore*, had been described by Ludwig Nohl as early as 1874: see his *Beethoven, Liszt, Wagner* (Vienna, 1874), pp. 79–80.

13. Thayer's volume seems to have come out in January 1879; since Nottebohm's rejoinder appeared in instalments in April and May, it must have been prepared very rapidly.

14. In attempting to reconstruct the Petter Sketchbook, he makes no comment on the fact that the first nine leaves have a completely different watermark from the rest.

15. Information kindly furnished by Dr. Rudolf Elvers.

16. In a somewhat later passage (1863, col. 399), in referring to the melodrama in the third act of the opera, Jahn even wrote of "die Skizzenbücher, in welchen dasselbe entworfen ist." He may have recognized this as an unguarded phrase, for in the revised (1866) version of his essay it was changed to: "die Entwürfe, welche sich unter den Skizzen finden."

17. I cannot accept H.-G. Klein's identification of the sketches on pages 345 and 346 as being for the *Leonore No. 3* overture (Klein, *Ludwig van Beethoven: Autographe und Abschriften*, pp. 231, 276, 277).

18. I am very grateful to Mr. Donald Greenfield for information on this point.

19. Nottebohm, *Zweite Beethoveniana* (p. 452n.), drew attention to the absence of any sketches here for the *Leonore No. 1* overture, Op. 138, which was in fact the third to be composed. For a recent discussion of the date of Op. 138, see my article referred to in note 2 above.

20. See Erich H. Müller, "Beethoven und Simrock," in *Simrock Jahrbuch* 2 (1929):27.

21. See Wilhelm Hitzig, "Aus den Briefen Griesingers an Breitkopf & Härtel entnommene Notizen über Beethoven," *Der Bär* (1927), p. 28.

22. Ibid., p. 30.

23. A letter of 2 October 1805 from Josef Sonnleithner, who prepared the German text of the libretto, to the theater censor includes the sentence: "Court-Secretary Josef Sonnleithner requests that the veto of 30 September this year on the opera *Fidelio* be lifted, on the grounds that . . . thirdly, Beethoven has spent over a year and a half on this composition." This seems pretty accurate. For the text of the letter, see Karl Glossy, "Beiträge zur Geschichte der Theater Wiens (1801–1820)," *Jahrbuch der Grillparzergesellschaft* 1 (1915): 83 ff.; reprinted in *Archiv für Musikwissenschaft* 2 (1920):404.

24. These sketchleaves were once owned by Joseph Joachim: see Sir George Grove's article on *Beethoven* in the first edition of his *Dictionary,* vol. 1 (1879), p. 184 n. 13.

25. See Wilhelm Virneisel, "Aus Beethovens Skizzenbüchern," *Colloquium Amicorum: J. Schmidt-Görg zum 70. Geburtstag* (Bonn, 1967), pp. 431–35.

26. When Beethoven used these four leaves, (c) was between fol. 1 and fol. 2 of (b).

27. The most extensive sketch is illustrated in *Die Flamme lodert: Beethoven-Ausstellung der Stadt Wien* (Vienna, 1970), p. 142.

28. Sketches for the first number on pp. 89–91, 146–47 and 189, and for the fifth number on pp. 92–93 and 95, represent revisions of earlier work.

29. This was "Copyist C," who also prepared the *Abschrift* of the "Eroica" Symphony in 1804. (For the identification, see Alan Tyson, "Notes on Five of Beethoven's Copyists," *Journal of the American Musicological Society* 23 (1970): 452–56 and plates 7 and 8.) The watermark evidence is as follows. Each number was copied by Copyist C on paper either with an "apple" watermark (an apple, or possibly a flaming heart, enclosing the letters *GA / F,* and a half-moon with a face) or with an "Emerich" watermark (the name G. A. EMERICH and three half-moons). Of these, the "apple" paper is clearly the earlier. The following numbers of the opera's first version are copied on "apple" paper: nos. 1 (C-major version), 4, 5, 6, 8, 9, 10, 14 (first version), 15. On "Emerich" paper are: nos. 1 (C-minor version), 11, 14 (second version), 16 and 18; the sketches for the *Leonore No. 2* overture in Landsberg 10, pages 21–28 (see H.-G. Klein, *Ludwig van Beethoven: Autographe und Abschriften,* p. 133) are also on this paper. It is easy to see from this that the opera was being copied number by number, as soon as each was composed; rewritten numbers were recopied on paper with the later watermark.

30. La Mara [Marie Lipsius], *Beethoven und die Brunsviks* (Leipzig, 1920), p. 59.

31. Joseph Schmidt-Görg, *Beethoven: dreizehn unbekannte Briefe an Josephine Gräfin Deym geb. v. Brunsvik* (Bonn, 1957), pp. 15–16; Harry Goldschmidt, *Um die Unsterbliche Geliebte* (Leipzig, 1977), p. 194 and note 551.

8

Paper as Evidence: The Utility of the Study of Paper for Seventeenth-Century English Literary Scholarship

William Proctor Williams

Ironically, the material upon which the literature of England was written and printed during the sixteenth and seventeenth centuries has been the least-exploited form of bibliographical and historical evidence. The missed opportunities because of this neglect, or apparent randomness of study, are terrible to contemplate. In what follows I intend to deal with the "state of the art" at the present time and then to demonstrate, from some recent research of my own, the importance of the study of paper.

It is one of the misfortunes of paper study that there seems to have always existed a sharp divide between those scholars who study paper as an artifact and those who study the writings placed upon it; in other words, between the students of paper history and the students of history, literature, and bibliography. This has been further complicated by the fact that the Newer Bibliography arose at the beginning of the present century as a response to the study of the texts of the English printed drama of the Renaissance; and that the parallel rise in the study of paper history has tended to concentrate on the work of paper mills, that is, industrial history, and primarily those on the continent, with little regard for the distribution of the products of the mills. However, for the period from the early eighteenth century onward, this divide has more and more been closed, and such instances as the long series of notes and articles by Alfred H. Shorter on British provincial papermaking and G. Thomas Tanselle's "The Bibliographical Description of Paper" (*Studies in Bibli-*

ography 24 (1971): 27–67) indicate how useful paper evidence has become.

From the publication of McKerrow's *Introduction to Bibliography* in 1927 and the subsequent establishment of the study of bibliography in universities, it has been normally thought sufficient for a scholar to know the difference between laid and wove paper, and enough of the paper's physical properties to determine printing formats by the position of watermarks and the direction of chain lines. Philip Gaskell's *New Introduction to Bibliography* (Oxford: Oxford University Press, 1972) moves things little further for the student of the pre-1700 period, though it should be noted that Gaskell is much more advanced on paper evidence, as on most other things, for the post-1700 era. Generally speaking, the scholar of the earlier period knows little about the potential of paper evidence and is not encouraged to press on farther.

The examples of W. W. Greg, A. W. Pollard, and, much later, Allan Stevenson on the paper used in the Pavier Quartos have not been regularly followed.[1] Many scholars have studied watermarks only when accident led them there, and it cannot be said that a thorough examination of paper evidence is always the first thing a scholar undertakes.

It must be admitted that in many sound critical editions of seventeenth-century works, the editors have taken care to fully examine all possible forms of evidence, including paper. But I must stress that the scrutiny of the paper of the literary document, printed or manuscript, is not yet assumed by all to be always as necessary as are the signings, type, format, and other such matters. Even if the scholar wishes to make an intensive study of the paper of the documents he is working with, his efforts will be, at best, blunted by the failure of the scholarly world to provide any comprehensive documentation of watermarks, paper supply, and the like, and he will be left to find the odd article on paper, to fall back on the old standards such as Briquet, Heawood, and Churchill, and then to call it a day. Much valuable work that has already been done in connection with a particular work or author is utterly inaccessible from the avenue of the subject "paper," and only that slippery research method known as "lucky chance" will be there to aid the scholar. Indeed, when two so influential scholars as Sir Walter Greg and Charlton Hinman, in their works on the Shakespeare First Folio, do not even find a place in the index of their books for "paper" or "watermarks," smaller fry can be excused for thinking that perhaps such things don't really matter after all.[2] It is further indicative that the *New Cambridge Bibliography of English Literature* (Cambridge: Cambridge University Press, 1974, 1:927–30) lists only fifty-six entries on the subject of paper and watermarks.

In part, such a situation may have arisen from the heavy concentration of work on printed materials, where it is assumed that matters of paper supply are so complex and so obscure as to be insoluble. Work on manuscripts, always a minority portion of textual study for this period, has been principally concerned with paleographical and textual matters.

However, hopeful signs are now to be seen. This present volume is certainly one of them, the recent publication of Irving P. Leif's *An International Sourcebook of Paper History* (New York: Archon, 1978) is another. The great increase in the study of book trade history will, I believe, go far toward remedying the supposed impossibility of uncovering details of paper supply for English printed books. And the gradual accumulation of radiographs of watermarks will eventually bring about a much greater concern for, and use of, paper as evidence not only of format and alterations in formats but also of dating, attribution, and, as I hope to show later here, at least rudimentary literary history and criticism.

The scholar making use of paper evidence must still begin his investigations with Briquet's *Les Filigranes* (edited by Allan Stevenson, 4 vols. [Amsterdam: Paper Publications Society, 1968]), W. A. Churchill's *Watermarks in Paper in Holland, England, France, etc. in the XVII and XVIII Centuries and Their Interconnection* (Amsterdam: Hertzberger, 1935), and Edward Heawood's *Historical Review of Watermarks* (Amsterdam: Swets & Zeitlinger, 1950) and *Watermarks* (Hilversum: The Paper Publications Society, 1950); E. J. Labarre's "The Study of Watermarks in Great Britain" (*The Briquet Album*, ed. E. J. Labarre, [Hilversum: The Paper Publications Society, 1952], pp. 97–106) provides a summary of work to about 1950. A good summary of the industrial side of paper study is provided by D. C. Coleman's *The British Paper Industry, 1495–1860* (Oxford: Clarendon Press, 1958).

Allan Stevenson's "Paper as Bibliographical Evidence (*Library*, 5th ser., 19 [1962]: 197–212) is a sound statement of methodology and a clear beacon to others that shows just what can be achieved; while Frank Sullivan's "Little Pitchers in the Big Years: Being a Study of the Water Pitcher Watermark in Elizabethan England" (*Paper Marker* 20 [1951]: 11–19), along with Stevenson's study of twin watermarks (*Studies in Bibliography* 4 [1951–2]: 57–91) indicate a trend that may be developed by scholars at the end of the twentieth century. However, before we can progress much far-ther, it will be necessary to have produced a complete bibliogra-phy of all bibliographical studies, both articles and books, that records all previously published work dealing in any way with paper evidence in descriptive bibliography and textual criticism. Such a bibliography would instantly advance the study of paper as evidence in the seventeenth century, for with it, we would be able to draw together the currently dispersed scholarship on the subject. Of course, it would also be useful if a comprehensive record of watermarks found in English documents, both printed and manuscript, could be produced. This would be both a long and expensive project, but with current advances in the copying of watermarks, and by publishing the results in some inexpensive form, such as microfiche, it seems possible to hope that some team of scholars might undertake such a project.[3] Finally, a union must be forged between those who work primarily from the history of industry and trade side of the subject and those whose main concern is the literary text. Book-trade historians have very much to tell the analytical bibliographer, and perhaps the ana-lytical bibliographer has something to tell the book-trade histo-rian. As long as we continue in the present semidarkness, neither will ever know.

I would now like to describe a specific example of the very great utility of paper as evidence in an instance dealing with seven-teenth-century English manuscripts. Although the example is probably not a typical one, its striking nature may perhaps serve to highlight how important it is for scholars to always extract every bit of information from the paper with which they are working.

In September 1977 a large collection of dramatic and poetic manuscripts from the seventeenth century was rediscovered at Castle Ashby, Northamptonshire, the home of the Marquess of Northampton. The collection had last been seen, and in-completely described, by Bishop Thomas Percy in 1767. Percy listed nine titles and ascribed them all to Cosmo Manuche, a dramatist of the mid-seventeenth century. However, the collec-tion actually contained thirteen titles, three of these existing in two drafts, as well as two short poems. These were contained in nine separate manuscript volumes; none of the works in this collection had ever been published.

The collection was subsequently purchased by the British Library, and the volumes and their contents are as follows: BL Addit. MS. 60273, "The Banished Shepherdess" by Cosmo Manuche, folio volume formed of mainly disjunct leaves; 60274, "The Feast" by Cosmo Manuche, folio volume formed of conjugate folds; 60275, "Love in Travell" by Cosmo Manuche, folio volume formed of conjugate folds; 60276, play on the subject of Saint Herminigildus, translations of Seneca's *Agamemnon* and *Hercules Furens*, two poems "The Cavaliers" and "Presbyterian," perhaps by James Compton, third earl of Northampton, small folio volume formed of six gatherings of quired sheets; 60277, translations of *Agamemnon* and *Hercules Furens* (fair copies of works in 60276) folio volume formed by quiring twenty-two folded sheets; 60278, translation of Machiavelli's *The Mandrake* and the first act of Corneille's *Don Sancho* and a fragment of "Leontius, King of Cyprus" (see also 60279), author and translator unknown, folio volume formed by quiring twenty folded sheets; 60279, "Leontius, King of Cyprus," author unknown, folio volume formed of conjugate folds; 60280, "Mariamne," author unknown but may be James Compton, third earl of Northampton, folio volume formed of conjugate folds; 60281, drama concerning the reign of the Emperor Caracalla, author unknown, folio volume formed of thirty-six sheets quired. Second copies of the texts found in 60273 and 60274 are located at the Huntington Library and Worcester College, Oxford, respectively.[4]

Only three of the works bear any contemporary assignment of authorship (the three plays by Manuche, "The Feast," "The Banished Shepherdess," and "Love in Travell," are all presentation copies with dedications signed by Manuche), and two important early questions were the date of composition of the various texts and the attribution of those titles that do not have an author. Although the plays varied widely in style and genre, they had one underlying tendency: to dwell upon the events of the English Civil War and the defeat of the royalist cause, either directly or allegorically. It was natural, therefore, to begin with the working assumption that all the manuscripts were closely associated in their composition with the royalist household of the earl of Northampton and that it would be likely that references to their authors, the plays themselves, or their performance, might be found among the family documents of the Comptons in the Muniments Room at Castle Ashby. Alas, no such evidence was discovered, but many of the estate record books from the 1660s were bound in the same format, with the same marbled paper covers as the literary manuscripts. For example, "The Book of Laborors Soms f[r]

Lady day last past . . . 1666" (Compton Family Document #1001) and several estate account books for the Ashby Estate for 1666 (CFD #1000) not only are in a similar format with similar covers but also are made up of paper bearing the same watermarks as the literary manuscripts. It would also appear that a great supply of these notebooks was on hand, perhaps from a restocking during the 1650s after the earl had compounded and regained possession of his properties, for such an account book is used to record deeds in 1700 (CFD #1246).

We may suppose that a large stock of these blank books would have been kept in the estate offices at both of the family's country homes, Castle Ashby in Northamptonshire and Compton Wyn-yates in Warwickshire, and probably at their temporary London residence, and that people of authority working on the estates, the family, and friends of the family might draw on this supply for various purposes. Although the Watermarks of the collection of literary manuscripts had initially been of importance only for the determination of format, one of the traditional uses of paper evidence, it now seemed possible to make the evidence of the paper yield up even more information.

It will be useful at this point to set out in a tabular form the significant watermarks and to show their appearance in all the literary manuscripts and other documents.

WM #1 (an armorial shield with a hunting horn hanging in the middle, 58 × 34 mm; like Churchill #s 313–31, especially #315)
"Banished Shepherdess" BL Addit. MS. 60273
"The Feast" BL Addit. MS. 60274
"Love in Travell" BL Addit. MS 60275
"Mandrake" and "Leontius" BL Addit. MS. 60278
Compton Family Documents (#1001, 1) 1666
Compton Family Documents (#1246) 1700
Compton Family Documents (#1083) 1660
MS. belonging to Isabella, Countess of Northampton, Library at Castle Ashby. 1660–1662
Letter by Earl of Northampton, BL Addit. MS. 29570 1644

WM #2 (three balls, one atop the other, with shield or crest on top of these, bars or ornaments in top two balls, 70 × 22 mm; similar to Heawood #320)
"Banished Shepherdess" BL Addit. MS. 60273
"The Feast" Worcester College, Oxford

WM #3 (an armorial shield with hunting horn hanging in the middle, the name "DVRAND" in compartment at foot, 65 × 50 mm; Heawood #1223)
"Mandrake." "Leontius" BL Addit. MS. 60278
Compton Family Documents (#1001) 1666
Compton Family Documents (#1083, 4) 1642
Compton Family Documents (/1083, 5) 1642
Compton Family Documents (#1083, 7) 1642
Compton Family Documents (#1083, 20) 1643
Compton Family Documents (#1083, 29) 1644

WM #4 (a pot with the letters IB or $\overset{P}{DB}$ in center—this is so common a watermark and the letters are so indistinct in most examples that all that can be said here is that there is a similarity)
"Agamemnon" BL Addit. MS. 60277
"Leontius" BL Addit. MS. 60279
"Mariamne" BL Addit. MS. 60280
"Caracalla" BL Addit. MS. 60281 (i)
Compton Family Documents (1083, 1) 1642
Note: The Huntington Library copy of "The Banished Shepherdess" has the watermark of a jester, or perhaps only a foolscap, and this watermark is not found in any of the other of the literary manuscripts or of the Compton family documents.

It will be immediately observed that at the very least, the evidence of the watermarks shows us that the manuscripts were not merely gifts from literary acquaintances or retainers of the Comptons, but that many of the manuscripts were written on the same paper that the Comptons were using regularly themselves for business records and personal and official correspondence. Thus, the literary manuscripts and their composition are tied rather closely to the household of James, the third earl of Northampton. Further, watermarks #1 and #3 allow for some greater precision in dating, since we know that these two kinds of paper were in earliest use by the household in 1642 (all instances of official business concerning the opening of the Civil War and some later military matters), and that such papers seem to have been most frequently used, particularly in the binding format found in the literary manuscripts, between 1660 and 1666. I believe that the use of paper with watermark #1 in 1700 is an aberration and that particular account book is an accidental survival. This is more likely since the watermarks in other documents from around 1700, which are in more plentiful supply in the family papers than for the period forty years earlier, all bear different and more "modern" watermarks.

However, the single thing that the watermark evidence supports is that all this active play writing (and it is very extensive, comprising everything from multiple translations of classical and continental Renaissance drama to political satires, to fully formed original plays, and it is concentrated into a period of about twenty years) was done so fully under the patronage of the earl and countess of Northampton that even the paper for composition was provided. And one would guess that the writing was done at Castle Ashby and that the manuscripts had never left the house until October 1977. On this point it is interesting to observe that the one copy that we know was always intended for another patron and location, the Huntington copy of "The Banished Shepherdess," which was a presentation copy for Queen Dowager Henrietta Maria, is written on paper unlike that found in any of the Castle Ashby manuscripts.

When one takes into account paleographical evidence that shows that James Compton, the third earl of Northampton, was the person who copied a large number of these manuscripts (BL Addit. MSS. 60276, 60280, and portions of 60277, 60278, 60279, and 60281) and that whether or not he is also the author of these works, he is certainly very much responsible for their revision and preservation, the corroborating evidence of the paper is most valuable. Although neither the paleographical nor the paper evidence by themselves would have been sufficient to more than tantalize, the two together make it clear that the Compton household employed itself during the time of the Commonwealth with writing and probably producing a large number of plays in a wide variety of styles for a courtly, royalist audience, a discovery that unaided literary criticism could never have made.

Notes

1. W. W. Greg, "On Certain False Dates in Shakespearian Quartos, *Library*, n.s. 9 (1908–9): 113–31 and 381–409; Alfred W. Pollard, *Shakespeare Folios and Quartos* (1909; reprint., New York: Cooper Square, 1970), pp. 81–107; Allan H. Stevenson, "Shakespearian Dated Watermarks," *Studies in Bibliography* 4 (1952): 159–64 and also "Watermarks are Twins," *Studies in Bibliography* 4 (1952): 57–91.

2. W. W. Greg, *The Shakespeare First Folio* (Oxford: Clarendon Press, 1953); and

Charlton Hinman, *The Printing and Proof Reading of the First Folio of Shakespeare*, 2 vols. (Oxford: Clarendon Press, 1963).

3. The work of Thomas L. Gravell and George Miller in their *A Catalogue of American Watermarks 1690–1835* (New York: Garland, 1979) shows the direction of work that I have in mind, but the qualifications to this particular work presented by David Schoonover in his review in *Papers of the Bibliographical Society of America* (74 [1980]: 283–85) indicate that much further work, and thought, must be given to such a task if we are to be given truly useful reference works on watermarks.

4. This collection was first described by me in *Times Literary Supplement* 9 (December 1977):1448, and a full description of the collection appeared in *Library*, 6th Ser., 2 (1980), 391–412.

9

The Analysis of Paper and Ink in Early Maps:
Opportunities and Realities

David Woodward

Within the last twenty years, several new means of analyzing the physical and chemical structure of historical artifacts have shown considerable promise. So rapid has been the development and so wide the choice of techniques available that their comparative value has become unclear, especially to practitioners in fields where analytical techniques have only recently been introduced. The aim of this essay is to compare some of the opportunities available and to pose some questions concerning their value for the analysis of early maps.

The logical analysis of physical form in printed books and manuscripts without the use of electronic aids has a much longer, if sporadic, history. In analytical bibliography, for example, the study of the Thomas Wise forgeries by Carter and Pollard in 1934 was one of the earliest attempts at using detailed physical evidence of paper and typography in conclusively demonstrating the falsity of documents.[1] Their conclusions were elegant in their logic and simplicity, and provided a methodological model by demonstrating the value of careful and systematic observation of physical detail, an aspect of study that had previously been neglected or even overlooked in favor of the document's content. In the history of cartography, this artifactual approach has already been summarized elsewhere by the author.[2] The recent addition of such techniques as beta-radiography, external beam particle-induced X-ray emission (PIXE), and energy dispersive X-ray fluorescence (XRF) has provided new opportunities for the analysis of both manuscript and printed maps that the historian of cartography should consider and that will be summarized here. The value of these techniques will depend on the advantages of each system for studying the matter at hand.

Opportunities

What are the physical components of a map that can be systematically analyzed? Simply expressed, they are the fabric (paper, vellum, etc.) and medium (ink, paint, etc.).[3] Some techniques, such as beta-radiography, are only applicable to paper analysis, while others may be applied to both the fabric and the medium.

Paper

The dating of paper used for maps has been of interest at least since the work of Edward Heawood, who combined a knowledge of paper history with that of the history of cartography, and who used maps as examples for his volume in the series of watermark albums published by the Paper Publications Society.[4] Heawood's interest in this evidence is also seen in a series of articles on maps printed in Italy in the sixteenth century.[5] Because most of these maps have complicated plate histories—plates often changed ownership several times during their lifespan—the paper evidence can be especially valuable.

Heawood's interest in watermarks and in sixteenth-century Italian printed maps was continued by the scholar-collector George H. Beans, who between 1957 and 1962 presented most of his map collection to the John Carter Brown Library.[6] In addition to his collecting, Beans published, in Jenkintown, Pennsylvania, an attractive series of publications under the imprint of the George H. Beans Library, that included his own and others' work and a small handlist of watermark tracings found on sixteenth-century Italian maps.[7] He also contributed to *Imago Mundi* on topics of his collecting interest under the title "Notes from the Tall Tree Library."[8] Further studies on the subject have been carried out by the present author since 1977.[9]

Comparative watermark analysis was severely hindered by the lack of an objective method of reproducing and recording the marks, a drawback that has now been largely solved by direct photography (Ilkley: a simple contact printing technique using a 15-watt light bulb and high-speed graphic arts film that is laid

FIGURE 1. Two watermarks in the Sotheby Atlas, ca. 1570. (Author's radiographs.)

under the map during exposure), ultraviolet (Dylux) and beta-radiography.[10] The hand tracing of watermarks, while acceptable for broad studies and certainly better than nothing, is now considered much less desirable than the objective methods.

In 1967, The Newberry Library acquired the Franco Novacco collection of approximately 330 sixteenth-century Italian maps, and between 1978 and 1980, the author acquired about 1,100 beta-radiograph images of watermarks from these and other sixteenth-century Italian printed maps that had been loaned to the library from other institutions and private collections. In addition, a set of watermark images was obtained from a sixteenth-century Italian atlas sold at Sotheby Parke-Bernet on 15 April 1980 to a private collector in England (its whereabouts are now unknown). Before the sale, the author catalogued a portion of the atlas in detail and photographed about one hundred ten watermarks using the Ilkley process. The process is fast and harmless to the document, but suffers from the drawback that the engraved detail is not omitted from the print and may sometimes seriously obscure the watermark.

The atlas consisted of a core of sixteenth-century maps inlaid in extended margins or marginal strips. The watermark evidence is crucial to establishing the date and place of assemblage, which was concluded to be Venice, c. 1570. The key marks, illustrated in figure 1, are the siren-in-circle and horse-in-circle, which are found respectively on the two sheets of the map of the world on a cordiform projection by Giovanni Paolo Cimerlino engraved in 1566. The siren watermark can be confidently identified as of Venetian origin, and its association with the horse mark on the Cimerlino map would suggest that the horse is also Venetian. This is the only map in the atlas in which the horse mark appears, but it occurs in the marginal strips with great frequency. One can therefore assume that the core of the atlas was assembled with the extended margins in a Venetian shop, probably in 1570, the date of the last map that has such margins.[11]

Further research projects at the University of Wisconsin have focused on all watermarks of one design from the entire collection of images. This design was a siren or mermaid with two tails in a circle surmounted by a star. Forty-eight watermarks representing thirty-seven maps were selected (some maps consisted of two or more sheets pasted together). Sixteen of these images were obtained from The Newberry Library, seven from Helsinki University Library, one from California State University, Fullerton, and twenty-four from private collections in California and London.

Forty-three of the images were beta-radiographs; the remaining were negatives made with the Ilkley process.

These forty-eight images were compared by eye and found to fall into two distinct groups, distinguished by a difference in the shape of the mermaid's right shoulder (assuming she is facing us). In the Martha-type watermark, the right shoulder was broader than the left. The other image was called Mary. Out of one thousand images taken randomly from several collections worldwide, only *two* paper moulds were represented, a surprising find. It indicated the likelihood that these two watermarks were from twin moulds, and therefore most probably were always used together in tandem in the papermaking process. This of course is not unusual in the making of handmade paper; but it does suggest that no other moulds using this emblem made paper on which maps were printed, an unexpected discovery.

Furthermore, five out of six sets of watermarks on the two-sheet maps turned out to be from the paired moulds, suggesting that, in a two-sheet map, there was a strong likelihood that the sheets were printed one after the other rather than that several copies of one sheet were run off and then several copies of the other. This conclusion assumes that sheets in a batch of paper (with inevitable exceptions) would normally be printed in the order they were made, and that the watermarks would follow an alternating pattern using the two different moulds in tandem.

Of the dates that were represented by the maps, the date range of the plates was 1559 to 1570. The frequency of the dating is shown in figure 2, and it can be seen that the frequency increases toward the latter part of the period. None was found dated after 1570. Beans gave a range of 1561 to 1570, but also found none after 1570.[12] Something happened to this pair of moulds in 1570, and the search is on for maps with such a mark bearing a publication date of 1571 or after. However, on the basis of the large sample already gathered, it is unlikely that such maps will be found. Although a small sample is illustrated in figure 2, it is possible to infer that the marks were current during the latter part of the period only (that is, from 1566 to 1570), and that earlier dated maps were simply printed from the earlier plates during those years.

The question arises: how do we know if the difference between two watermarks is due to two states of the same mould or two different moulds? Fortunately, a technical detail in the manufacture of the mould comes to our aid. The watermark was usually attached to the mould with thin sewing wires that show up as dark dots on the radiograph. Even if the shape of the mark should

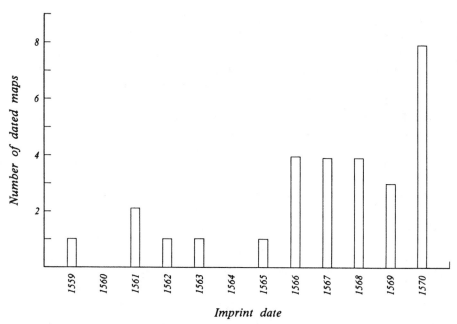

FIGURE 2. Frequency of dated maps, 1559–1570. *(Courtesy of Wisconsin Cartographic Laboratory.)*

become distorted with use, therefore, the patterns of the sewing dots will remain the same (figure 3). On the other hand, the likelihood of two marks on different paper moulds having the same pattern of sewing dots is slim indeed. In some cases, it is true, a sewing dot might be added to strengthen the mould, but this will aid and not hinder the ordering of the states of the mould.

A confident identification of the paper moulds in this experiment could not have been made before the techniques of watermark photography and radiography had been developed. But the possibilities for analysis certainly do not end there. Stevenson has shown that paper moulds may go through identifiable stages in their lives that are analogous to the stages of a printing plate. The mark may become increasingly distorted with use as the mould is jostled or as excess pulp is brushed from it. More dramatically, if the mark is situated between chain lines and not sewn to a chain line passing through it, the sewing wires tend to become loose with age and the mark moves slowly to the left in relation to the chain lines (figure 4). Stevenson estimated the rate of movement as averaging about a millimeter a month, a distance certainly discernible on a radiograph.[13]

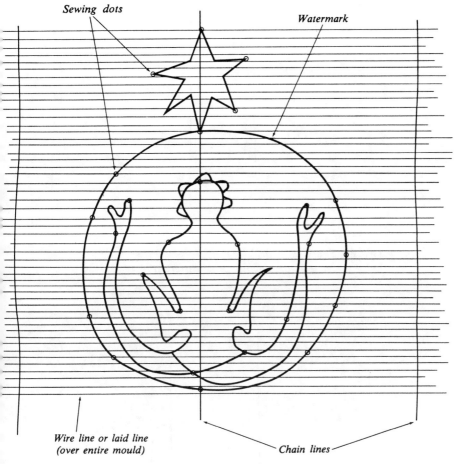

Sewing dots

Watermark

Wire line or laid line
(over entire mould)

Chain lines

FIGURE 3. Pattern of sewing dots on watermark. *(Courtesy of Wisconsin Cartographic Laboratory.)*

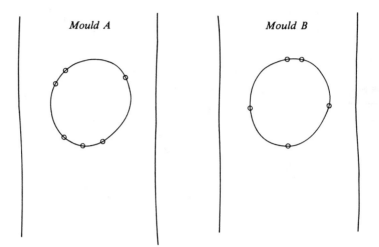

Twins = Shape similar;
Pattern of sewing dots different

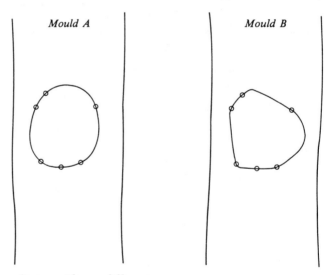

States = Shape different;
Pattern of sewing dots similar

FIGURE 4. Movement of watermark in relation to chain lines. *(Courtesy of Wisconsin Cartographic Laboratory.)*

This theory is promising, but there are practical difficulties. The sewing dots are not always perfectly distinguishable, even on the radiograph. Further, since each image has to be compared with every other image to discern minute differences, the number of combinations exceeds eleven hundred. With this in mind, it was decided to take thirty-nine of the forty-eight images (those already in film form) to the University of Wisconsin's Center for Remote Sensing, which recently acquired equipment for the analysis of satellite imagery, particularly Landsat.[14] These images were converted to numerical form on a scanning microdensitometer that records the film density of the radiograph at each of 350,000 small squares, here shown at normal size and enlarged eight times (figures 5 and 6). For each square, or picture element, three pieces of information are stored on tape or disk: the x and y coordinates of the picture element and the recorded density.[15]

Once the images are in digital form, they can be manipulated statistically in several ways. The range of density can be standardized from image to image by stretching the contrast between a given low and high figure. Further, the contrast of the images may be enhanced to bring out the pattern of sewing dots. If two images of watermarks from the same mould are superimposed on an image processor, these dots will become more prominent. If they are from different moulds, this will also become immediately apparent.

The analysis of successive states of a watermark using beta-radiography can most easily be achieved when the mark is not wired to a central chain line (thus allowing it to slide along the wire lines during its lifetime). This was the basis of Stevenson's study. But for watermarks that are tied to a central chain line—such as in the case of the vast majority of sixteenth-century Italian marks—analysis of this clarity is not feasible, since the variation of the position of the mark between the chain lines is usually not great enough to be measured.

Nevertheless, the stresses placed on the paper mould during its life sometimes cause the shape of the watermark to change subtly, and a continuation of the study using the thirty-nine siren watermarks has shown that minute changes in the shape of a mark can also be recognized using a combination of precise measurement and statistical analysis. The coordinates of twelve control points were chosen on each of the thirty-nine watermarks from the digital image displayed on the image processor, normalized using an affine transformation, and the root mean square (RMS) error that measured how well one set of control points fitted another

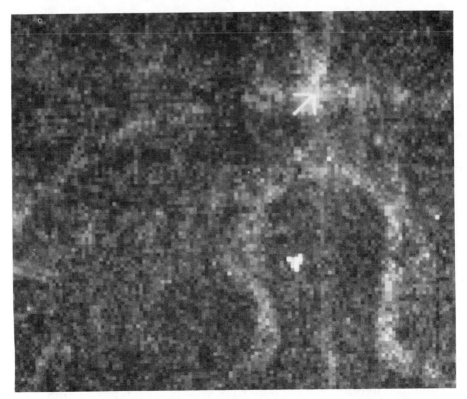

FIGURE 5. Images of sewing dots from a scanning microdensitometer that records the film density of the radiograph at each of 350,000 small squares, here shown at normal size. *(Author's photograph.)*

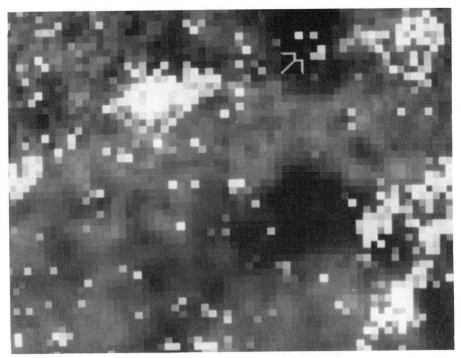

FIGURE 6. The image shown in figure 5, here enlarged eight times. *(Author's photograph.)*

was computed. These were tabulated in a matrix so that the fit of every watermark on every other watermark within the same mould could be readily seen. (It was found earlier that the RMS values could readily distinguish between two marks from different paper moulds; these values were much higher than for those from the same mould).The radiographs of the best and worst fit cases were then carefully examined to establish where the extreme differences lay. In the case of the Martha mould, the extremely subtle difference was seen in the degree of roundness of the left (as we see it) shoulder (figure 7). The same is true of the Mary mould, but the angle of the *V* between the fin and right shoulder changes very slightly (figure 8). It is hypothesized that, over the life of a mould, those curved wires subject to horizontal pressure in the brushing off of excess pulp from the mould at the end of the day would become increasingly angular. Correspondingly, we would expect the most curved examples to be the earlier states of the mould. No map on either Martha or Mary paper has yet been found bearing a date after 1570, so we may postulate this as being the end of the mould's life. Of the dated maps bearing the Martha watermark, 1559 is the earliest, yet the state of this mark is similar to that of a map dated 1569. If we accept Stevenson's view that paper stocks of normal sizes were used up quite quickly, say within one year, it would seem likely that the life of the Martha (and thus probably also the Mary) mould might be reasonably postulated to be between 1568 and 1570, or perhaps even 1569–1570. This conforms with the evidence presented earlier in this essay that toward the end of the decade we see a marked increase in the number of dated maps bearing marks from the Martha and Mary moulds, thus considerably narrowing the range given by Beans (1561–1570) and providing a more precise tool than was previously thought.

PIXE

Beta-radiography has now been joined by techniques that measure the percentages of elements in paper. For example, Particle-Induced X-Ray Emission, or PIXE, has now been used successfully in archeological and bibliographical work (in addition to its more usual biological and chemical applications).[16] A beam of protons is accelerated in a cyclotron, deflected into a vacuum pipe, and narrowed down to a precise beam that can be made less than a millimeter square to be aimed at the document in question. In order to avoid placing the document in a vacuum, the beam is

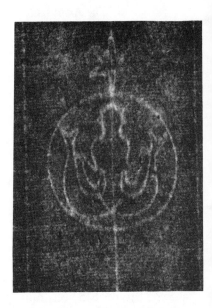

FIGURE 7. Radiographs of the best and worst fit cases of the Martha mould. *(Author's photograph.)*

FIGURE 8. Radiographs of the best and worst fit cases of the Mary mould. *(Author's photograph.)*

passed into a helium or air chamber into which the document is introduced. This improvement, known as "external beam," is essential for the handling of large, awkwardly shaped, or precious artifacts, including maps and atlases. When aimed at a section of a document, either at the paper, vellum, ink, or pigment, the clashing of particles in the beam with the atoms of the various elements in the object being analyzed excites the atoms in such a way as to generate characteristic X rays that shoot out in all directions. A sample of these is read and the characteristic X rays of each element present in the section of document under analysis are counted, processed by computer, and recorded. In order to avoid bias due to different thicknesses of the material analyzed, the occurrence of an element is expressed as a ratio to calcium, which is a common element in paper of any age.[17]

Each sheet of paper seems to have its unique chemical profile, and the technique is so sensitive that in a study of an eighteenth-century octavo French travel book undertaken by a team at the University of California, Davis, the signatures were revealed as groups of eight relatively homogeneous leaves.[18] The sensitivity of this technique is underlined by the fact that the physicist who drew attention to the periodicity was not previously aware of the occurrence of signatures in printed books.

In a more recent study reported by Eldred, 324 leaves from the first volume of a Gutenberg Bible from St. John's Seminary, Camarillo, California, were analyzed using the UC Davis cyclotron. Calcium again was found to be the most abundant element, with smaller amounts of silicon, phosphorus, potassium, sulfur, iron, manganese, copper, and zinc. By various combinations of iron and manganese in three categories of watermarked paper, it was possible to determine the category of an unwatermarked page from the chemical analysis alone.[19]

The implications of this technique for the study of map paper are several; for the present purpose, examples in the study of sixteenth-century Italian printed maps have been chosen. For these maps, many of which have been extracted from composite atlases with marginal strips pasted to their borders, the technique could easily be used for reconstructing the original content of these atlases by grouping marginal strips of similar chemical composition. In addition, the PIXE data could be used to answer a series of questions about the chemical variation of papers with the same watermark, its twin, or its variant of the same design. Papers bearing the same watermark, taken from several different types of document, such as printed books or prints, could also be analyzed. Perhaps most important, papers of similar chemical content with different watermarks might be searched for in order to establish the association of multiple watermark designs using a common papermaker's vat at various times. Finally, were enough data gathered, what Schwab has called a "systematic chemical-bibliographical grid" could be compiled for a given period, and samples of unknown origin placed within it.[20]

The main problem with the analysis of paper is that it dates the paper and not the impression. While Stevenson attempted to allay our fears about this, skepticism remains. With enough data on the chemical composition of the paper, however, along with analysis of the other physical component of the document—the ink—both interpreted within their general publishing context, it might indeed be possible to arrive at a good estimate for the average shelf life of a sheet of paper between paper mould and printing press, and thus an indication of the precision by which impressions may be dated from an analysis of the paper that carries them.

Ink

Unlike the analysis of paper, the analysis of printing ink on maps provides information about the circumstances of the impression and printing rather than of the papermaking, and is thus a more directly useful form of evidence. Yet if the history of paper is an obscure area of study, the study of printing ink as a historical source of evidence is far more arcane, largely because the methods of analyzing it have not been available until very recently.[21] One reason given for the delay in study of the Vinland map inks in the late 1960s, for example, was that improvements in microspectroscopy had to be awaited before the analysis could be completed. The more recent PIXE or XRF techniques, which we have already discussed in relationship to paper analysis, promise to revolutionize this field.

The studies now being carried out using PIXE at the University of California, Davis, with the Gutenberg Bible reveal an astonishing sensitivity of the technique in analyzing the composition of printing ink. By analyzing the ink of a single character on the verso of each of the 324 leaves, as well as testing several samples from the same page, Eldred was able to show independently that Gutenberg printed the book one page at a time rather than one folio at a time, and that four printing presses were used at various stages of the work; two to start with, which were then joined by two others. Moreover, the pages printed by each press can be identified.[22]

A technique capable of providing conclusions of such technical detail can clearly also add an important dimension to the studies of map paper. In the study of sixteenth-century Italian printed maps, for example, key ratios of the composition of ink could be plotted against key ratios of the composition of paper, and the resulting clusters would indicate which certain combinations were particularly active. Should these clusters also be related to the printing of certain map plates or particular centers of the map trade (for example, Venice), further conclusions could be drawn. Composite atlases suspected of being partially printed at once, such as a Venetian atlas in the Newberry Library previously described by the author, could be analyzed with this method to confirm this idea.[23]

XRF

Competing with the PIXE technique is Energy Dispersive X-Ray Fluorescence (XRF), which has been in wide use in analytical chemistry since about 1950 and has been used in the study of archeological and fine arts objects for several years.[24] Both wavelength and energy dispersive systems have been used, but only the latter may be nondestructive. In the energy dispersive system, an X-ray beam is focused on a thin surface layer of the sample, which fluoresces in all directions, producing electrical pulses whose magnitude is unique to each element present in the sample. A detector close to the source senses some of these pulses, which are counted and processed, usually by a microcomputer, to provide proportions of elements above Sodium (atomic number 11). Some units have different ranges depending on whether the sample is placed in air or helium.

At the Winterthur Museum in Wilmington, Delaware, studies of paper and graphic objects have been carried out with success since 1973, enabling the staff to obtain recognizable spectral patterns for papers from specific paper mills made over several decades. The technique is even valuable for nineteenth-century artifacts: by detecting the presence of zinc or copper uniformly distributed over a lithographic print, the composition of the printing plate can be deduced. Lithographs free of zinc or copper are thus assumed to have been printed from stone, thus predating the lithographic transfers from metallic plates. The application of this to the recognition of late nineteenth-century states of lithographic maps is clearly promising.[25]

More recently, Gary Carriveau of the Detroit Institute of Arts used XRF for an analysis of the pigments on several Rembrandt drawings and found that a recent unrecorded restoration on one of them had been carried out using a pigment containing titanium dioxide.[26] The Detroit XRF equipment was also used in a study by Bèla Nagy, who analyzed several pigments on selected European maps from the fifteenth to the nineteenth centuries, clearly demonstrating the value of the technique for detecting modern color. For example, on a 1681 map of Lombardy by Cantelli da Vignola, a number of nineteenth- and twentieth-century pigments were found, namely zinc white, titanium white, and barium white.[27]

Realities

Given the several options available to the researcher who wishes to gather physical evidence from early maps, it is important to consider several factors in the choice of method, and to be aware of possible improvements that are likely to take place in the near future. In particular, the question must be faced whether the analysis of elements in paper and ink by the PIXE technique, which can identify individual batches of paper and ink with great sensitivity, has rendered other analytical techniques obsolescent.

One group of factors is certainly the availability, mobility, and expense of the equipment. Cyclotrons are not mobile, and not every university, let alone researcher, has access to one, particularly one that has been adapted for historical artifacts. Cahill listed the eleven PIXE research programs that were using an external beam in the world in 1980, including four in the United States and one in Great Britain.[28] While this number has no doubt increased since that report, the expense of obtaining the use of a cyclotron already heavily committed to biological, medical, and environmental research is obviously high. Add to this the problems of transporting valuable artifacts to it, and it becomes clear that only a large well-funded program of research could feasibly use such facilities. Nevertheless, the interest shown in bibliographical questions by the research team at the Crocker Nuclear Laboratory at the University of California, Davis, demonstrates that, with the cooperation of lending institutions, significant work can be done at a reasonable cost. XRF analysis is more available, since the equipment is much smaller and the initial capital cost is less. Some units may also be transportable to the artifact. In addition, the cost of the analysis is inexpensive and highly trained personnel are not usually required. But the number of units that have modifications for the analysis of historical artifacts is again severely limited.[29]

Equally as important as the availability of the equipment are the concerns of curators and librarians in protecting historical artifacts from irreversible damage. According to Cahill, no technique other than external beam PIXE and XRF seems to fulfill this essential requirement.[30] Chemical or electron beam methods, such as the X-ray diffraction and electron microscopy used for the Vinland map, which have necessitated destroying samples of the artifact, however small, would now appear to be less desirable.[31]

Another group of factors includes the technical requirements of the analysis, such as size of the target area, sensitivity of the system, errors caused by unevenness in the surface of the sample, and range of elements detectable. PIXE can detect, in principle, all elements between sodium and uranium in a single irradiation. Depending on the unit, XRF can only detect about thirty of these elements above chlorine (atomic number 17), although recent models with the sample placed in helium can detect above sodium. The accuracy of PIXE (normally ± 5 or ± 2 percent for thin targets) also exceeds that of XRF in equivalent irradiation time. The unevenness of the surface of the sample has an effect on accuracy in both XRF and in PIXE. We must await further experience with the analysis of historical papers and inks to determine the level of sensitivity required, although Hanson reports that XRF provides data well within the needs of the Winterthur staff for the purpose of detecting forgeries in general museum artifacts.[32]

The main advantage of PIXE over XRF at present seems to be in the size of the target analyzed at equivalent times of irradiation. PIXE can focus to 1 mm in an exposure of thirty seconds, but this would take much longer for XRF. Exposure time was about 5 minutes for a 5 mm diameter target area in the Nagy study, but in order for this to be focused to 1 mm, the irradiation time would have to be about 125 minutes. The sensitivity desired must therefore be weighed against the time and expense of the analysis. Since the composition of the sample is averaged over the target area, more sensitive readings will result from a smaller target area. In the analysis of maps, this is a prime consideration, as the ink is frequently found only on very thin lines. Pigments may also be found confined in small areas. For a general analysis of large areas of paper or pigment, however, XRF may be adequate for the purposes of the research at hand. The sensitivity and viewing area of the XRF technique is, however, rapidly improving, and it might well provide a valuable alternative to the PIXE technique, particularly for cases where larger areas can be sampled or where the precision requirements are not as stringent.

If external beam PIXE can identify a batch or even a sheet of paper with a unique chemical fingerprint, to say nothing of the ink, what future is there for watermark analysis? On the surface, it might appear that all current projects for compiling albums of watermark images should be discontinued in favor of systematic PIXE or XRF analysis of whole groups of documents from various periods and origins. But it is equally desirable to compile files of watermark images, preferably using prints from beta-radiography negatives produced at full size. The reason is that for many purposes, such as the determination of forgeries, a dating precision of only a few years may be necessary. In addition, for the analysis of an occasional suspect document, a quick beta-radiograph or other watermark image is more feasible than an individual analysis by XRF or PIXE, even if the latter were available locally.

Furthermore, the systematic collection of watermark images flags those documents that are suitable candidates for PIXE or XRF analysis. For example, a PIXE analysis of all thirty-nine of the sixteenth-century Italian siren watermarks in the author's recent work, together with additional similar marks gathered from books and prints, would demonstrate the variation in composition of the batches created with this particular pair of watermarks. This might demonstrate that only a sample of papers of a given watermark might need PIXE analysis. Should the ink in maps, books, and prints of particular printers be found to correlate with batches of paper bearing such watermarks, further confirmation might also be found of the author's narrowing of the date of appearance of maps appearing on this paper to 1568 to 1570.

The main limiting factor of all scientific methods of analysis at present is that a sufficient fund of characteristic data has yet to be built up. The information relating to the chemical content of paper or ink means little in isolation: it needs to be related to the norms for a particular period, printer, or papermaker. Considerable institutional cooperation will be necessary if this information is to be gathered systematically and in a consistent format. A videodisk archive is one possibility. Furthermore, if beta-radiograph images were stored digitally, they would be accessible by telecommunications. Statistical data relating to the content of paper, ink, and pigment should also be made available in digital form. Despite the apparent immensity of the task, it is not too early to start to compile specifications for such a data archive, which, if coordinated by a major library or institution, would constitute an impressive resource, not only for historians of cartography, but for all researchers, conservators, archivists, librarians, and others who need access to precise physical information about the documents that come into their hands.

Notes

The assistance of the John Simon Guggenheim Memorial Foundation and the Graduate School of the University of Wisconsin, Madison, is gratefully acknowledged.

1. John Carter and Graham Pollard, *An Enquiry into the Nature of Certain Nineteenth-century Pamphlets* (New York: Scribners, 1934).

2. David Woodward, "The Form of Maps: An Introductory Framework," *Antiquarian Bookman Yearbook*, (1976), pp. 11–20.

3. While they are not strictly physical components, the marks themselves (ink lines, patches of color) have physical characteristics that can be measured to provide clues to the origin or dating of documents. For example, microphotography of the impressions has been used to order states of Hogarth prints, and enhancement of writing on manuscripts has been carried out by Benton et al. See John F. Benton, Alan R. Gillespie, and James M. Soha, "Digital Image-Processing Applied to the Photography of Manuscripts," *Scriptorium* 19 (1979): 40–55. Additional physical components might include adventitious matter, such as dirt, stains, etc., some of which may reveal the history of a particular document. Arthur Baynes-Cope has summarized the physical components of documents in "The Scientific Examination of the Vinland Map at the Research Laboratory of the British Museum," *Geographical Journal* 140 (1974): 208–11. Additional terms for these components (none of which has yet entirely been accepted in the literature) are discussed in David Woodward, "The Form of Maps: An Introductory Framework."

4. Edward Heawood, *Watermarks, Mainly of the 17th and 18th Centuries* (Hilversum: Paper Publications Society, 1950).

5. Edward Heawood, "The Use of Watermarks in Dating Old Maps and Documents," *Geographical Journal* 63 (1924): 391–412; "Italian Map Collections of the Sixteenth Century," *Geographical Journal* 69 (1927): 598–99; "An Undescribed Lafreri Atlas and Contemporary Venetian Collections, *Geographical Journal* 73 (1929): 359–69; "Another Lafreri Atlas," *Geographical Journal* 80 (1932): 521–22; and "Another Lafreri Atlas," *Geographical Journal* 87 (1936): 92–93.

6. John Carter Brown Library, "The George H. Beans Gift of Maps and Geographical Treatises," Brown University, John Carter Brown Library, *Report*, 1957, pp. 14–33; 1958, pp. 46–59; 1959, pp. 33–39; 1960, pp. 31–33; 1960–65, p. 45.

7. George H. Beans, *Some Sixteenth Century Watermarks Found in Maps Prevalent in the "IATO" Atlases* (Jenkintown, Penn.: George H. Beans Library, 1938).

8. George H. Beans, "Notes From the Tall Tree Library," *Imago Mundi* 4 (1947): 24; 6 (1949): 31; 7 (1950): 89; 8 (1951): 15; 9 (1952): 108; 10 (1953): 14; 11 (1954): 146; 12 (1955): 57; 13 (1956): 163; 14 (1959): 112; 15 (1960): 119, and 16 (1962): 160.

9. David Woodward, "The Study of the Italian Map Trade in the Sixteenth Century: Needs and Opportunities," *Wolfenbütteler Forschungen* 7 (1980): 137–46; "New Bibliographical Approaches to the History of Sixteenth-Century Italian Map Publishing" (Paper delivered at the Eighth International Conference on the History of Cartography, Berlin, 1979); "New Tools for the Study of Watermarks on Sixteenth-Century Italian Printed Maps: Beta Radiography and Scanning Densitometry," in Carlo Clivio Marzoli, ed., *Imago et Mensura Mundi: Atti del IX Congresso Internazionale di Storia della Cartografia*, 2 vols., (Rome: Istituto della Enciclopedia Italiana, 1985), 2:541–52.

10. The techniques of reproducing watermarks, with their various advantages and disadvantages, are discussed in David Schoonover, "Techniques of Reproducing Watermarks" in this anthology. A useful summary and history of the use of beta-radiography for bibliographical purposes can be found in G. Thomas Tanselle, "The Bibliographical Description of Paper," *Studies in Bibliography* 24 (1971): 27–67, especially pp. 48–50. Reproductions of radiographs from maps are found in R. V. Tooley, "Notes and Addenda, *"Map Collectors' Series* 34 (1967), plate 4; and in Richard W. Stephenson, "The Delineation of a Grand Plan," *Quarterly Journal of the Library of Congress* 36 (1979): 207–24, photograph p. 215. The work of Thomas Gravell with Dylux ultraviolet techniques has also included some map subjects. See Thomas Gravell, "Watermarks and What they Can Tell Us," in *Preservation of Paper and Textiles of Artistic Value*, vol. 2, ed. John C. Williams (Washington, D.C.: American Chemical Society, 1981), pp. 57–62, and his manual of American watermarks, *Catalogue of American Watermarks, 1690–1835* (New York: Garland Press, 1979).

11. Sotheby Parke Bernet & Co., *Catalogue of Highly Important Maps and Atlases* [including the] *Doria Atlas of Sixteenth Century Italian Maps*, sale date, 15 April 1980 (London: Sotheby Parke Bernet & Co., 1980), pp. 32–33.

12. George H. Beans, *Some Sixteenth Century Watermarks*.

13. Allan Stevenson, *The Problem of the Missale Speciale* (London: The Bibliographical Society, 1967), pp. 248–52.

14. The assistance of Frank Scarpace, Pete Weiler, and Mark Olsen is gratefully acknowledged.

15. Scanning was performed on an Optronics Photomation System P-1700 scanning microdensitometer with a resolution of 500 × 700 picture elements. The images were processed on a Stanford 170 Digital Image Processor.

16. For a general discussion of PIXE, see T. A. Cahill, "Proton Microprobes and Particle-Induced X-Ray Analytical Systems," in J. D. Jackson, H. E. Gove, and R. F. Schwittens, eds., *Annual Review of Nuclear Particle Science* 30 (1980): 211–52.

17. T. A. Cahill, B. Kusko, and R. N. Schwab, "Analyses of Inks and Papers in Historical Documents Through External Beam PIXE Techniques," *Nuclear Instruments and Methods* 181 (1981): 206.

18. Richard N. Schwab, "The Cyclotron and Descriptive Bibliography: A Progress Report on the Crocker Historical and Archaeological Project at UC Davis," *Book Club of California. The Quarterly News-Letter* 47 (1981): 3–12.

19. Robert A. Eldred, "External Beam PIXE Programs at the University of California, Davis," Institute for Electrical Engineering and Electronics, *Transactions on Nuclear Science* 30 (1983): 1276–79

20. Richard N. Schwab, "The Cyclotron and Descriptive Bibliography," p. 9.

21. See, however, the excellent bibliography in Monique De Pas and Françoise Flieder, "Historique et perspectives d'analyse des encres noires manuscrites," in International Council of Museums, Committee for Conservation, Report of the meeting in Madrid, 2–7 October 1972, 5 pts. in 3 vols. (Rome: International Center for Conservation, 1972).

22. Robert A. Eldred, "External Beam PIXE Programs," pp. 1276–79.

23. David Woodward, "New Tools for the Study of Watermarks."

24. For a general study, see E. P. Bertin, *Introduction to X-Ray Spectrometric Analysis* (New York: Plenum Press, 1978).

25. Victor F. Hanson, "Determination of Trace Elements in Paper by Energy Dispersive X-ray Fluorescence," in John C. Williams, ed., *Preservation of Paper and Textiles of Artistic Value*, 2:147–48.

26. Gary W. Carriveau and Marjorie Shelley, "A Study of Rembrandt Drawings Using X-Ray Fluorescence Analysis," *Nuclear Instruments and Methods* 193 (1982): 297–301.

27. Bèla Nagy, "The Colormetric Development of European Cartography" (M.S. thesis, Eastern Michigan University, 1983).

28. T. A. Cahill, "Proton Microprobes," pp. 211–52.

29. Bèla Nagy, "The Colormetric Development of European Cartography," pp. 58–91.

30. Robert A. Eldred, "External Beam PIXE Programs," p. 1276.

31. For a brief summary of the techniques used on the Vinland Map, see Walter C. McCrone and Lucy B. McCrone, "The Vinland Map Ink," *Geographical Journal* 140 (1974): 212–14.

32. Victor. F. Hanson, "Determination of Trace Elements," p. 145.

Select Bibliography

Alibaux, Henri. *Les premières papeteries francaises.* Paris, 1926.

Alston, Robin. "Reproducing Watermarks." *Direction Line* 2 (1976): 1–3.

Balston, Thomas. *William Balston, Paper Maker, 1759–1849.* London, 1954. Reprint. New York, 1979.

———. *James Whatman, Father and Son.* London, 1957. Reprint. New York, 1979.

Barker, Nicolas, and John Collins. *A Sequel to An Enquiry into the Nature of Certain Nineteenth Century Pamphlets.* London and Berkeley, 1983. See Carter, John.

Barnes, Warner. "Film Experimentation in Beta-Radiography." *Direction Line* 1 (1975): 3–4.

Bidwell, John. "Paper and Papermaking: 100 Sources." *AB Bookman's Weekly* (13 February 1978): 1043–61.

Blum, André. *On the Origin of Paper.* Translated by Harry Miller Lydenberg. New York, 1934.

Bofarull y Sans, Francisco de. *Animals in Watermarks.* Hilversum: Paper Publications Society, 1959.

———. *Heraldic Watermarks; or, La Heráldica en la Filigrana del Papel.* Hilversum: Paper Publications Society, 1956.

Briquet Album, The. See Labarre, Emile Joseph.

Briquet, Charles Moise. *Briquet's Opuscula: The Complete Works of Dr. C. M. Briquet without Les Filigranes.* Hilversum: Paper Publications Society, 1955.

———. *Les Filigranes: Dictionnaire historique des marques du papier.* Geneva, 1907. Reprint. Leipzig, 1923. *The New Briquet—Jubilee Edition.* Edited by Allan Stevenson. 4 vols. Amsterdam: Paper Publications Society, 1968.

Browning, B. L. *Analysis of Paper.* New York, 1969.

Bühler, Curt F., "The Margins in Medieval Books." *Papers of the Bibliographical Society of America* 40 (1946): 34–42.

———. "Watermarks and the Dates of Fifteenth-Century Books." *Studies in Bibliography* 9 (1957): 217–24.

———. "Last Words on Watermarks." *Papers of the Bibliographical Society of America* 67 (1973): 1–16.

Carter, John, and Graham Pollard. *An Enquiry into the Nature of Certain Nineteenth Century Pamphlets.* London, 1934. Reprint. New York, 1971.

Edited with an epilogue by Nicolas Barker and John Collins. London, 1983.

Chapman, Robert W. *Cancels*. London and New York, 1930.

Churchill, William Algernon. *Watermarks in Paper in Holland, England, France, etc., in the XVII and XVIII Centuries and their Interconnection*. Amsterdam, 1935. Reprint. Amsterdam, 1967.

Clapperton, Robert Henderson. *Modern Paper-Making*. London, 1929. 3d ed. Oxford, 1952.

———. *Paper, An Historical Account of Its Making by Hand from the Earliest Times Down to the Present Day*. Oxford, 1934.

———. *The Paper-Making Machine: Its Invention, Evolution, and Development*. Oxford, London, New York, 1967.

Coleman, Donald C. *The British Paper Industry, 1495–1860*. Oxford, 1958.

Collins, John. See Barker, Nicolas, and Carter, John.

Eineder, Georg. *The Ancient Paper-Mills of the Former Austro-Hungarian Empire and their Watermarks*. Hilversum: Paper Publications Society, 1960.

Frost, Kate. "Supplement to Leif: A Checklist of Watermark History, Production, and Research." *Direction Line* 8 (1979): 33–56.

Gaskell, Philip. "Notes on Eighteenth-Century British Paper." *Library*, 5th ser., 12 (1957): 34–42.

———. *A New Introduction to Bibliography*. Oxford, 1972.

Gerardy, Theo. "Die Wasserzeichen des mit Gutenbergs kleiner Psaltertype gedruckten Missale speciale." *Papiergeschichte* 10 (1960): 13–22.

———. "Zur Datierung des mit Gutenbergs kleiner Psaltertype gedruckten Missale speciale." *Archiv für Geschichte des Buchwesens* 5 (1963), cols. 399–415.

———. *Datieren mit Hilfe von Wasserzeichen*. Bückeburg, 1964.

———. "Wann wurde das Catholicon mit der Schluss-Schrift von 1460 (GW 3182) wirklich gedruckt?" *Gutenberg Jahrbuch* 1973: 105–25.

———. "Die Datierung zweier Drucke in der Catholicontype (H 1425 und H 5803)." *Gutenberg Jahrbuch* 1980: 30–37. See also Pulsiano #143.

Gravell, Thomas L. "A New Method of Reproducing Watermarks for Study." *Restaurator* 2 (1975): 95–104.

———. "The Need for Detailed Watermark Research."*Restaurator* 4 (1980): 221–26.
See also Pulsiano # 152–54 and Schoonover, pp. 162–63 in the present volume.

Greg, Sir Walter W. "On Certain False Dates in Shakespearian Quartos." *Library*, 2d ser., 9 (1908): 113–31, 381–409.

Hazen, Alan T. *A Bibliography of the Strawberry Hill Press*. New Haven, 1942.

Heawood, Edward. "The Position on the Sheet of Early Watermarks." *Library*, 4th ser., 9 (1928–29): 38–47.

———. *Watermarks, Mainly of the 17th and 18th Centuries*. Hilversum: Paper Publications Society, 1950.

———. *Historical Review of Watermarks*. Amsterdam, 1950. See Pulsiano #165–71, Introduction, p. 23 n.15, and Woodward, p. 219 n.5, in the present volume.

Hellinga, Lotte, and Hartel, Helmar, editors. *Buch und Text im 15. Jahrhundert*. Hamburg, 1981.

Herdeg, Walter, ed. *Art in the Watermark*. Zurich, 1952.

Hudson, Frederick. "The New Bedford Manuscript Part-Books of Handel's Setting of *L'Allegro*." *Notes: The Quarterly Journal of the Music Library Association* 33 (March 1977): 531–52. See Pulsiano #204–6 and Hudson, pp. 45–46, 58 n.14 in the present volume.

Hunter, Dard. *Old Papermaking*. Chillicothe, Ohio, 1923.

———. *Papermaking through Eighteen Centuries*. New York, 1930. Reprint. New York, 1970.

———. *Papermaking: The History and Technique of an Ancient Craft*. New York, 1943. 2d rev. ed. New York, 1947.

———. *Papermaking by Hand in America*. Chillicothe, Ohio, 1950.

———. *Papermaking in Pioneer America*. Philadelphia, 1952. Reprint. 1981. See Pulsiano #208–18.

Janot, Jean Marie. *Les Moulins à Papier de la Région Vosgienne*. 2 vols. Nancy, 1952.

Johnson, Douglas, and Alan Tyson. "Reconstructing Beethoven's Sketchbooks." *Journal of the American Musicological Society* 25 (1972): 137–56. See also Pulsiano #232 and Hudson, pp. 46–47 in the present volume.

Labarre, Emile Joseph. *Dictionary and Encyclopaedia of Paper and Paper-Making, with Equivalents of the Technical Terms in French, German, Dutch, Italian, Spanish and Swedish*. Rev. ed. Amsterdam, 1952. Supplement by E. G. Loeber, 1967.

Labarre, Emile Joseph, ed. *The Briquet Album*. Hilversum: Paper Publications Society, 1952; Chapel Hill, 1981.

———. "The Study of Watermarks in Great Britain." In *The Briquet Album*, pp. 97–106. Hilversum: Paper Publications Society, 1952.

———. *A Short Guide to Books on Watermarks*. Hilversum: Paper Publications Society, 1955. See Pulsiano #267–71.

Lalande, Joseph Jérôme le Français de. *Art de faire le papier*. Paris, 1761. 2d ed. Neuchatel, 1776. Published separately Paris, 1820.

LaRue, Jan. "Watermarks and Musicology." *Acta Musicologica* 33 (1961): 120–46.

LaRue, Jan, and J. S. G. Simmons, "Watermarks." *New Grove Dictionary of Music and Musicians* 20. London, Washington, D.C., Hong Kong, 1980. See also Hudson, pp. 37–39, in the present volume.

Le Clert, Louis. *Le Papier: Recherches et notes pour servir à l'histoire du papier, principalement à Troyes et aux environs depuis le quatorzième siècle.* Paris, 1926.

Lehrs, Max. *Geschichte und kritischer Katalog des deutschen, niederländischen und französischen Kupferstichs im XV. Jahrhundert.* Vienna, 1908–34.

Leif, Irving P. *An International Sourcebook of Paper History.* Hamden, Conn., and Folkestone, Kent, 1978.

Lindt, Johann. *The Paper-Mills of Berne and their Watermarks, 1465–1859.* Hilversum: Paper Publications Society, 1964.

Loeber, E. G. *Paper Mould and Mouldmaker.* Amsterdam: Paper Publications Society, 1982. See Labarre, Emile Joseph.

McKerrow, R. B. *An Introduction to Bibliography for Literary Students.* Oxford, 1927.

Meder, Joseph. *Dürer-Katalog.* Vienna, 1932. Reprint. New York, 1971.

Mosin, Vladimir. *Anchor Watermarks.* Amsterdam: Paper Publications Society, 1973.

Needham, Paul. "Johann Gutenberg and the Catholicon Press." *Papers of the Bibliographical Society of America* 76 (1982): 395–456.

Nicolai, Alexandre. *Histoire des moulins à papier du sud-ouest de la France, 1300–1800: Périgord, Agenais, Angoumois, Soule, Béarn.* Bordeaux, 1935.

The Nostitz Papers: Notes on Watermarks Found in the German Imperial Archives of the 17th and 18th Centuries, and Essays Showing the Evolution of a Number of Watermarks. Hilversum: Paper Publications Society, 1956.

Papermaking, Art and Craft. Washington: Library of Congress, 1968.

Piccard, Gerhard. "Die Datierung des Missale speciale (Constantiense) durch seine Papiermarken." *Archiv für Geschichte des Buchwesens* 2 (1960): 571–84. See Pulsiano #339–59 for references to Piccard's *Findbücher*, etc.

Pollard, Graham. "Notes on the Size of the Sheet." *Library,* 4th ser., 22 (1941–42): 105–37. See also Carter, John.

Povey, Kenneth, and I. J. C. Foster. "Turned Chain Lines." *Library,* 5th ser., 5 (1950–51): 184–200.

Schlosser, Leonard B. *An Exhibition of Books on Papermaking: A Selection of Books from the Collection of Leonard B. Schlosser.* Philadelphia, 1968.

Schulte, Alfred. "Papiermühlen-und Wasserzeichenforschung." *Gutenberg Jahrbuch* (1934): 9–27.
See Pulsiano #401–16.

Shorter, Alfred H. *Paper Mills and Paper Makers in England, 1495–1800.* Hilversum: Paper Publications Society, 1957.

———. *Paper Making in the British Isles: An Historical and Geographical Study.* Newton Abbot, 1971; New York, 1972.

Simmons, J. S. G. "The Leningrad Method of Watermark Reproduction." *Book Collector* 10 (1961): 329–30. See LaRue, Jan.

Smith, David Clayton. *History of Papermaking in the United States, 1691–1969.* New York, 1970.

Sotheby, Samuel Leigh. *The Typography of the Fifteenth Century.* London, 1845.

———. *Principia typographica.* London, 1858.

Spector, Stephen. "Symmetry in Watermark Sequences." *Studies in Bibliography* 31 (1978): 162–78.

Spicer, A. D. *The Paper Trade.* London, 1907.

Stevenson, Allan. H. "Watermarks Are Twins." *Studies in Bibliography* 4 (1951–52): 57–91.

———. "Chain-Indentations in Paper as Evidence." *Studies in Bibliography* 6 (1954): 181–95.

———. "Briquet and the Future of Paper Studies." *Briquet's Opuscula,* Hilversum: Paper Publications Society, 1955.

———. *Catalogue of Botanical Books in the Collection of Rachel McMasters Miller Hunt.* Vol. 2. Pittsburgh, 1961.

———. *Observations on Paper as Evidence.* Lawrence, Kansas, 1961.

———. "Paper as Bibliographical Evidence." *Library,* 5th ser., 17 (1962): 197–212.

———. *The Problem of the Missale speciale.* London, 1967. See Briquet, Charles Moise; also see Pulsiano # 451–58 and Introduction, p. 24n.27 in the present volume.

Strength and Other Characteristics of Book Paper 1800–1899. Richmond: W. J. Barrow Research Laboratory, 1967.

Sutermeister, Edwin. *The Story of Papermaking.* Boston, 1954.

Tanselle, G. Thomas. "The Bibliographical Description of Paper." *Studies in Bibliography* 24 (1971): 27–67.

Thomson, Alistair G. *The Paper Industry in Scotland, 1590–1861.* Edinburgh, 1974.

Tschudin, Walter Friedrich. *The Ancient Paper-Mills of Basle and Their Marks.* Hilversum: Paper Publications Society, 1958.

Tyson, Alan. "The Problem of Bethoven's 'First' Leonore Overture." *Journal of the American Musicological Society* 28 (1975): 292–334. See Johnson, Douglas; also see Pulsiano # 483–85 and Hudson, pp. 47–52 in the present volume.

Valls i Subirà, Oriol. *Paper and Watermarks in Catalonia.* Amsterdam: Paper Publications Society, 1970.

Voorn, Henk. *The Paper Mills of Denmark & Norway and Their Watermarks.* Hilversum: Paper Publications Society, 1959.

———. *De papiermolens in de provincie Noord-Holland.* Haarlem, 1960.

———. *Old Ream Wrappers: An Essay on Early Ream Wrappers of Antiquarian Interest.* North Hills, Penn., 1969.

———. *De papiermolens in de provincie Zuid-Holland, alsmede in Zeeland, Utrecht, Noord-Brabant, Groningen, Friesland, Drenthe.* The Hague, 1973.

Weeks, Lyman Horace. *A History of Paper-Manufacturing in the United States, 1690–1916.* New York, 1916. Reprint. New York, 1969.

Weiss, K. T. *Handbuch der Wasserzeichenkunde.* Leipzig, 1962.

Ziesche, Eva, and Dierk Schnitger. "Elektronenradiographische Untersuchungen der Wasserzeichen des Mainzer Catholican von 1460." *Archiv für Geschichte des Buchwesens* 21 (1980), cols. 1303–50.

Zonghi, Aurelio. *Zonghi's Watermarks.* Hilversum: Paper Publications Society, 1953. See Pulsiano # 523–25.

Notes on Contributors

CURT F. BÜHLER (1905–1985) was Research Fellow for Texts Emeritus at the Pierpont Morgan Library. His publications include *The Fifteenth Century Book, Early Books and Manuscripts,* and many other books and journal articles.

FREDERICK HUDSON was Reader in Music, University of Newcastle upon Tyne, until his retirement in 1978. His many journal articles include studies of Handel and Bach, and he has edited *J. S. Bach: Trauungskantaten,* and *G. F. Handel: Sechs Concerti Grossi.*

HILTON KELLIHER read English at Oxford before becoming an Assistant Keeper of Manuscripts in the British Library. He is the author of the exhibition catalogue *Andrew Marvel: Poet & Politician,* and of essays and articles relating mainly to Renaissance English and Anglo-Latin literature.

JOHN NÁDAS is on the faculty of the Department of Music at the University of North Carolina at Chapel Hill. His publications include essays on medieval music manuscripts and on Verdi.

PHILLIP PULSIANO is an Assistant Professor at Villanova University. His publications include *Bárðar saga Snæfellsáss* (edited and translated with Jón Skaptason) and articles on Old English language and literature.

DAVID SCHOONOVER is Curator of the American Literature Collection at the Beinecke Rare Book and Manuscript Library, Yale University. He has worked extensively with methods of watermark reproduction.

ALAN TYSON is a Senior Research Fellow of All Souls College, Oxford. His research is centered mainly on Beethoven and

Mozart. For many years he has been concerned with the evidential value of paper studies in investigating the dates of musical manuscripts, and their structures.

WILLIAM PROCTOR WILLIAMS, Professor of English at Northern Illinois University, is the editor of *Analytical & Enumerative Bibliography*. He has published a descriptive bibliography of Jeremy Taylor to 1700, a checklist of Jeremy Taylor from 1700 to 1976, the index to the Stationers' Register 1640–1708, and many articles, monographs, and notes on sixteenth- and seventeenth-century English literature, bibliography, and textual criticism.

DAVID WOODWARD is Professor of Geography at the University of Wisconsin at Madison, Director of the University Cartographic Laboratory, and Project Director of the History of Cartography Project. He is presently engaged in a six-volume, multi-author worldwide history of cartography from prehistory to the present day.

Index

Page references that appear in boldface type indicate illustrations.